APE INTO HUMAN

APE INTO HUMAN

A Study of Human Evolution

SECOND EDITION

S. L. WASHBURN

University of California, Berkeley

RUTH MOORE

Little, Brown and Company

BOSTON TORONTO

LIBRARY OF CONGRESS CATALOG CARD NO. 79-92803

FIRST PRINTING

Published simultaneously in Canada
by Little, Brown & Company (Canada) Limited

PRINTED IN THE UNITED STATES OF AMERICA

THE FIRST EDITION OF THIS BOOK WAS ENTITLED
APE INTO MAN: A STUDY OF HUMAN EVOLUTION.

CREDITS FOR ILLUSTRATIONS

The sources of the illustrations appear below. Many pieces have been redrawn. The artists are Gloria Pelatowski (Figures 1.4, 1.6, 1.7, 1.8, 1.9, 1.10, 5.1, 5.2); V. Susan Fox (Figures 2.3, 2.9, 2.10, 3.2, 3.3, 3.5, 3.8, 5.8, and 6.4); Richard H. Sanderson (Figures 4.2, 6.2); and Eric Stoelting (Figures 2.1, 2.5, and 2.6). The authors wish to thank the publishers, authors, photographers, and illustrators for granting permission to use their material. The figures without specified credits have been drawn for this book.

Frontispiece: Tim White.

Chapter 1: 1.1 From Franz Weidenreich, *The Skull of Sinanthropus Pekinensis: A Comparative Study on a Primitive Hominid Skull* (Lancaster, Pennsylvania: Lancaster Press, 1943). 1.2 Tim White. 1.3 Courtesy Field Museum of Natural History, Chicago. 1.4 From Raoul E. Benveniste and George J. Todaro, "Evolution of Type C Viral Genes: Evidence for an Asian Origin of Man," *Nature* 261 (13 may 1976), pp. 101-108. Adapted and reprinted by permission. 1.5 Oregon Regional Primate Research Center, William Montagna, Director. 1.6 From Vincent M. Sarich and John A. Cronin, "Molecular Systematics of the Primates," in Morris Goodman and Richard E. Tashian, eds., *Molecular Anthropology* (New York: Plenum Publishing Co., 1975), pp. 139-168. Adapted and reprinted by permission. 1.9 and 1.10 From "The Breakup of Pangaea" by Robert S. Dietz and John C. Holden, *Scientific American*, October 1970. Copyright © 1970 by Scientific American, Inc. All rights reserved.

Chapter 2: 2.2 Brian O'Connor. 2.3 Adapted with permission of Macmillan Publishing Co., Inc. from *Primate Evolution* by Elwyn L.

Simons, Fig. 12, p. 155. Copyright © 1972 by Elwyn L. Simons. 2.7 From *Historia Naturalis* by Ulysses Aldrovandus 1642. 2.8 From *In the Shadow of Man*. Copyright © 1971 by Hugo and Jane van Lawick-Goodall. Reprinted by permission of Houghton Mifflin Company. 2.9 Drawn from a photograph in Jane van Lawick-Goodall, *In the Shadow of Man* (Boston: Houghton Mifflin, 1971); copyright © 1971 by Hugo and Jane van Lawick-Goodall. 2.10 Drawn from a photograph in *Scientific American* 217 (December 1967): 29, by permission of Lee Boltin.

Chapter 3: 3.1 S. L. Washburn. 3.2 Drawn from a photograph by Ralph Morse in *The Primates;* © 1965 Time Inc. 3.4 Phyllis Dolhinow. 3.6 From Jane van Lawick-Goodall, *In the Shadow of Man*. Copyright © 1971 by Hugo and Jane van Lawick-Goodall. Reprinted by permission of Houghton Mifflin Company. 3.7 From R. V. S. Wright, "Imitative Learning of a Flaked Stone Technology—The Case of an Orangutan," in S. L. Washburn and E. R. McCown, eds., *Human Evolution: Biosocial Perspectives,* Perspectives on Human Evolution series, vol. IV (Menlo Park, California: Benjamin/Cummings, 1978), pp. 214, 220, 221. Reprinted by permission. 3.8 Adapted from W. E. Le Gros Clark, *Fossil Evidence for Human Evolution* (Chicago: The University of Chicago Press); copyright © 1955 The University of Chicago.

Chapter 4: 4.1 Jerry Cooke, *LIFE* Magazine; © 1949 Time Inc. 4.3 Adapted from W. E. Le Gros Clark, *Fossil Evidence for Human Evolution* (Chicago: The University of Chicago Press); copyright © 1955 The University of Chicago. 4.4 J. Desmond Clark. 4.5 and 4.6 From "Hominid Fossils from the Area East of Lake Rudolf, Kenya: Photographs and a Commentary on Context" by Richard E. Leakey and Glynn Ll. Isaac, in *Perspectives on Human Evolution 2,* edited by S. L. Washburn and Phyllis Dolhinow. Copyright © 1972 by Holt, Rinehart and Winston, Inc. Reproduced by permission of Holt, Rinehart and Winston, Inc. 4.7 Tim White 4.8 and 4.9 F. Clark Howell 4.10 and 4.11 Phillip V. Tobias.

Chapter 5: 5.3 From *Prehistory of Africa* by J. Desmond Clark (London: Thames and Hudson, 1970), by permission of the publisher. 5.4, 5.5, 5.6, and 5.7 S. L. Washburn. 5.8 Adapted from *The Primates:* © 1965 Time Inc. 5.9, 5.10, and 5.11 From Jane van Lawick-Goodall, *In the Shadow of Man*. Copyright © 1971 by Hugo and Jane van Lawick-Goodall. Reprinted by permission of Houghton Mifflin Company.

Chapter 6: From *Scientific American,* September, 1978, p. 206. Copyright © 1978 by Scientific American, Inc. All rights reserved. 6.2 From R. D. Lockhart, G. F. Hamilton, and F. W. Fyfe, *Anatomy of the Human Body* (1959), adapted and reprinted by permission of Faber & Faber, Ltd. 6.3. From Phillip V. Tobias, "The Cranium and Maxillary Dentition of *Australopithecus (Zinjanthropus) Boisei,*" in L. S. B. Leakey, ed., *Olduvai Gorge,* vol. II (Cambridge: Cambridge University Press, 1967). Reprinted by permission. 6.4 Reprinted by permission of Times Books, a division of Quadrangle/The New York Times Book Co., Inc., from W. E. Le Gros Clark, *The Antecedents of Man: An Introduction to the Evolution of Primates*. Copyright © 1959 by

W. E. Le Gros Clark. 6.5 From Adolph H. Schultz, "Illustrations of the Relation between Primate Ontogeny and Phylogeny," in S. L. Washburn and E. R. McCown, eds., *Human Evolution: Biosocial Perspectives,* Perspectives on Human Evolution series, vol. IV (Menlo Park, California: Benjamin/Cummings, 1978), pp. 255-283. Reprinted by permission. 6.6 and 6.7 Reprinted with permission of Macmillan Publishing Co., Inc. from *The Cerebral Cortex of Man* by Wilder Penfield and Theodore Rasmussen. Copyright © 1950 by Macmillan Publishing Co., Inc. 6.8 S. L. Washburn 6.9 From *The Anatomy of a Gorilla,* The Henry Cashier Raven Memorial Volume, American Museum of Natural History and Columbia University (New York: Columbia University Press, 1950), p. 102. Reprinted by permission. 6.10 and 6.11 From Wilder Penfield and Lamar Roberts, Speech and Brain Mechanisms (Princeton, New Jersey: Princeton University Press, 1959), pp. 131, 130. Reprinted by permission.

PREFACE

This book, intended to be useful in a wide range of anthropology courses, provides a glimpse of the changing view of human evolution. The main sources for that view are discoveries and datings of fossils and stone tools that help us describe the past; field studies of living primates that tell us what our nearest relatives are like; and advances in molecular biology and immunochemistry that illuminate our relationships to other primates, past and present.

These were the sources for the first edition of this book also, but since 1974 new research in each area has settled some controversies and opened others. Our picture of human evolution is in many ways vastly different and based on far stronger methods of proof. This new edition may help clarify these new developments by stating what we have learned and by pointing to new work now in progress.

As before, we have included some history in order to put the changing views of evolution into perspective, but we have tried to avoid discussing issues that are now only of historical interest. To simplify somewhat, we have stressed the contributions of a few people. We have tried, however, to select those contributions in a way that tells of the complex intellectual environment in which they are made. As before, also, we stress behaviors, because the course of evolution is determined by natural selection—that is, by the behaviors that led to survival.

This book is based on conversations between the authors and the joint use of many sources. One of us (S. L. Washburn) is responsible for the scientific point of view, and the other (Ruth Moore) for the mode of presentation. We wish to thank Alice Davis for major editorial assistance in preparing the manuscript. We are indebted to several people for useful advice and discussion: David Howard Day, Monroe Community College; Phyllis Dolhinow, University of California, Berkeley; Betty Goerke, College of Marin; Robert S. O. Harding, University of Pennsylvania; Donald M. Valdes, Denison University; and Laura Greer Vick, University of North Carolina, Chapel Hill. Our thanks also to the Alfred P. Sloan Foundation for contributing to the completion of this work.

CONTENTS

APE INTO HUMAN

1

OUR BEGINNINGS

As life spread out of the seas and onto the land and into the air, countless millions of living forms evolved. More than 1.5 million species are believed to live in the world today. Some of this multitude are as tiny as the virus; some are as huge as the whale; some, like the cheetah, have achieved great speed; others have eagle vision; some have persisted through more than a billion years, whereas innumerable others have perished; the millions of the present species inhabit the mountains, the plains, and the depths of the earth; the arctics and the jungles; the deserts and the wetlands.

All species have many successful attributes for life on this small, varied planet, but of the untold billions of individuals and the vast number of species of the past and present only one species evolved our large type of brain, our kind of upright posture, our gift for language, our self-consciousness, and the human way of life. Only one animal became human—and only in the last 4 to 8 million years did this happen. That even one should have made the transition to the human form was long considered as inconceivable and impossible by scholars.

Former Views

Until 1859 nearly all believed, often as an article of religious faith, that the one who was so different from all the others was a special creation, a fair, vernal being who had been set down in the Garden of Eden at seven o'clock in the morning in the year 4004 B.C. Lucas Cranach the Elder painted the idyllic scene—Adam and Eve, perfect and unsullied, standing beneath the laden apple tree while the gentle creatures of the earth—lambs, deer, and dogs—gamboled at their feet.

This ordained view of human origin was abruptly challenged on November 24, 1859, when Charles Darwin published his *Origin of Species.*[1] Although Darwin made only one guarded reference to man as such, he demonstrated with nearly incontrovertible thoroughness that all life had descended from one beginning, and ultimately from one kind of primordial cell. The implication was inescapable that humans, as part of nature, had descended from some earlier, unquestionably animal stock. Angry and shocked critics reacted in opposition and charged Darwin with proposing that apes were man's ancestors. Twelve years later, in *The Descent of Man,*[1] Darwin faced the question and said outright that "man is an offshoot of the Old World Simian stem."

The assertion that humans, with their special standing and uniqueness, were descended from crude, hairy creatures swinging in the trees of Africa, outraged much of the world. Darwin was accused of "brutalizing humanity and sinking mankind to a lower grade of degradation than any into which it has fallen since its written records tell us of its history."[2]

Darwin could answer only that humans ought to admit their community of descent frankly: "It is only our natural prejudice and arrogance which made our forefathers declare that they were descended from demi-gods." Even those who were willing to listen demanded proof. There stood the apes—here stood humans; obviously they were different. If they were related, where were the forms between them? Where were the missing links?

A number of years before Darwin's controversial works were published, workmen digging in the Neander River valley in Germany, in 1856, had come upon part of an exceedingly odd skull—the heavy brow ridges and thick cranial bones were unlike those of any living person. Rudolf Virchow, a leading pathologist of the nineteenth century, held that the skull was an abnormal specimen. It had belonged, another expert declared, "to an individual affected with idiocy and rickets." Darwin took no position, but his friend and defender, Thomas Huxley, made a thorough study of the puzzling skull and noted: "We meet with ape-like characters, stamping it as the most pithecoid [apelike] skull yet discovered." He hastened to caution: "In no sense can the Neanderthal bones be regarded as the remains of a human being intermediate between men and apes."[3] The controversy was enough, however, to lead the German scientist Ernst Heinrich Haeckel to hypothesize that a half-ape, half-human creature might have existed somewhere, sometime, and to suggest that if it should ever be found, an appropriate name would be *Pithecanthropus erectus* (upright apeman).

A Dutch physician, Eugène Dubois, whose imagination had been fired by the whole issue, was able to get a government appointment to Sumatra in order to be in a position to take up the search. He reasoned that the chance of further discoveries might be good in an area where the orangutan—in Malay the name meant "forest man"—still survived. After some years, in 1890, on the banks of a Java river Dubois found the fossil his scientific imagination had visualized—a low, apelike skull; and nearby, seemingly belonging to the same being, he unearthed a humanlike thigh bone. "I considered it a link between apes and men," said Dubois, and he named his discovery *Pithecanthropus erectus*. For those unwilling to admit to any relation between apes and humans, it was raw provocation, and the denunciations that followed were furious.

In 1927 Davidson Black, a Canadian physician, found similar fossils in ancient cave deposits in China, near Peking. (See Figure 1.1.) Along with the bones were skillfully made tools and hearths on which these early human hunters—not in-between creatures—had cooked their game.

Nevertheless, skepticism had met the announcement in 1925 that a true "missing link"—a creature

Figure 1.1 Skulls of female gorilla, Peking fossil, and *Homo sapiens*. These pictures symbolize the change of ape into human.

with a skull very like that of an ape, yet with human-like teeth—had been found in South Africa by Raymond Dart, a South African anatomist, who discovered it when he opened a box of fossilized bones sent to him from a quarry at Taung. The skull, that of a six-year-old, was very small, no higher than that of a living ape, but with essentially human teeth. He named the find *Australopithecus* (*Australo* for south, *pithecus* for ape). Scientists were the dubious and the rejectors this time, but Dart and Robert Broom, a Scottish physician who had gone to South Africa to search for fossils, continued their work.

Find succeeded find, not only in South Africa, but also in East Africa, where Mary and Louis Leakey explored the fossil-rich deposits of Olduvai Gorge, in Tanzania. From lakeside campsites of some millions of years ago came adult skulls, a foot, a hand, hundreds of teeth, and thousands of stones chipped into simple tools for cutting and perhaps for killing. Exacting studies were made of all this yield from the past. The confirmation was complete and beyond all scientific doubt —the African creatures being unearthed with brains no larger than those of the apes had been bipedal; they were upright and walked like humans.

In the summer of 1978 another startling discovery was made. At Laetoli, about twenty miles south of Olduvai Gorge, Mary Leakery and her co-workers uncovered human footprints in some ancient rock.[4] When the rock had been a soft, ashy stretch of ground, two persons and numerous animals had trekked across it. The tracks were as clear as though they had been made yesterday (see Figure 1.2). The two beings, one larger than the other, walked along, leaving a print at each step. The smaller one apparently paused and, after a little turn to the left—perhaps to look at something—continued on in the same direction as the other. In the second season of work the tracks were followed for seventy-five feet. Soon after the tracks were made, ash from a nearby volcano gently settled down into the de-

Figure 1.2 Human footprints found at Laetoli, Tanzania, by Mary Leakey. The impressions were made in soft volcanic ash that later solidified. The footprints have been filled with black sand to bring out the pattern. The makers toed out and the big toe was in line with the others. There is nothing remarkable about these human footprints—except that they are 3.6 million years old!

pressions covering the footprints and thus preserving them for millennia to come. When they were uncovered and the volcanic deposits above and below them were dated by radioactive means, specifically by the potassium-argon method, it was revealed that they were made an astounding 3.6 million years ago. The footprints, nothing like those of apes, could have been made only by creatures who walked upright. To test what their eyes told them, the scientists walked across soft earth. The impress of their feet was almost indistinguishable from the 3.6-million-year-old footprints. The correctness of their great age was reconfirmed when fossilized teeth and jawbone fragments were found in the same area.

At about the same time the Leakeys' footprint discovery was made, a research team working in the Afar region of Ethiopia unearthed the pelvis of another early australopithecine, a pelvis that went with an upright posture and that differed decidedly from the pelvis of a quadrupedal ape. It was dated at 3 million years and showed as clearly as the footprints at Laetoli that this australopithecine also was a biped.

Neither site yielded any of the chipped stone tools that were often found in abundance at other australopithecine locales. The same absence of tools had long puzzled scientists who had worked at the Sterkfontein site in South Africa, where the first adult australopithecine skull had been found in 1936. It was not until 1979 that Sterkfontein was found to be more than 3 million years old; the fossils had occurred in mixed cave deposits that had long seemed impossible to date. Finally, however, a strong effort accomplished the impossible, and the great age of Sterkfontein was disclosed.

The record thus became consistent. By all the available evidence the earliest, most primitive australopithecines, with their ape-sized brains of about 450 cc, had not mastered the art of stone-tool making. Tools were found only with australopithecine fossils dating back 2.5 million years or less. Although the hands of the upright early forms were free, it apparently took about a million years to learn to chip a few bits from a fist-sized piece of stone and to use its edge for hunting or other purposes.

Had the footprints, the Afar pelvis, and the date of Sterkfontein been discovered when the first australopithecine skulls were found in the 1920s, much misunderstanding and controversy could have been averted. It would have been clear that the earliest australopithecines had branched off from their ape ancestors a million or so years prior to 3.6 million years ago—probably between 4 and 8 million years in the past.

Nor would there have been confusion about the evolution that changed ape into hominid. It again would have been clear that bipedalism came first and that the other changes that would turn ape into modern human—tool use, larger brains, and the human way of life—were to come much later, well after the 3.6 million years of some of the earliest of humans, the primitive australopithecines.

In 120 years the evidence the world had demanded of Darwin had been partly accumulated.[5] The record went back to times more remote than almost anyone had deemed possible, and many missing links had been found. With the aid of the fossils, we could see what we had been like along most of the way: some 3 to 4 million years ago, an unexpected and odd combination with a brain no larger than an ape's and a humanlike body; about 1.5 million years ago, a larger-brained human who made complex stone tools. The record was still fragmentary, but it was complete enough to prove that humans had evolved. The famous French scientist Georges Cuvier once had proclaimed: "L'homme fossile n'existe pas"—there's no such thing as fossil man. He and most of the world had been wrong.

Before the Australopithecines

Other questions then arose. If the australopithecines were to be considered human, were they the first in this category? What kind of creatures had preceded them? From what stock did they spring?

The answers were far from easy to supply. From the time of the earliest australopithecines, more than 3 to about 8 million years ago, there was a dearth of fossils. The gap was a troublesome one. Prior to 8 to 10 million years, only the fossilized teeth and bones of apes were to be found. They were fairly numerous but were fragmentary. All of the material was meticulously studied. But because some of the fossils older than 8 million years displayed a few characteristics that were not wholly apelike, students debated about whether a tooth or a bit of skull was more apelike or more human.

Each bone fragment was compared with similar fragments of other apes, australopithecines, and humans. However careful these studies of comparative anatomy were, interpretations and opinions still differed. No one could say with scientific certainty that one or another of the known groups of changing apes was ancestral to australopithecines and humans.

It was possible that our ancestors sprang from some still earlier primate, and some investigators asked whether the separation occurred, not in the range of 4 to 8 million years, but perhaps as much as 50 million years ago. Under the circumstances, some scholars turned away from the Darwin-Huxley theory that humankind had descended directly from some earlier ape. Many maintained then, and some still do, that the human line separated far in the past, perhaps about 50 million years ago. Others made a strong case for a separation from some quadrupedal monkey about 30 million years ago, or from one of the early apes dating back about 20 million years.

When did the separation occur? If the human line branched off only 4 to 8 million years ago, the ape almost certainly was the ancestor rather than a quadrupedal monkey or the much more primitive tarsier. The recent dating of the earliest australopithecines—creatures just a step beyond the apes—at 3.6 million virtually settled the controversy.

The important date, however, was arrived at quite late—in 1978. For many years anthropologists also had to ask if any of the possible ancestral animals, whether tarsier or monkey or early ape, manifested any beginning traces of the behaviors that were later identified as uniquely human. Were any of them evolving a nascent bipedal locomotion, tool use, intelligence, or language?

For more than a hundred years it was generally believed that when the apes left the trees and descended to the ground, the most bipedal—those best able to walk upright—were the ones to survive and leave similarly gifted descendants. This theory was destroyed when new studies of apes in the wild showed that the quadrupedal chimpanzees and gorillas spend most of their time on the ground, and although they are not bipedal they survive very well. They can stand or take a few steps upright, but when they move for more than a short distance they "knuckle walk," using their bent-under fingers as forelegs. The question then became: How did ground-living, knuckle-walking apes evolve into bipedal humans, free to use the hands in all manner of activity? (See Figure 1.3.)

Questions always multiplied. The series of fossil discoveries had supplied evidence of the various stages of our evolution. Comparative anatomy studies further clarified many points of relationships. Radioactive dating resolved speculation about many dates, and studies of animals in the wild threw light on the evolution of behaviors. But there was still no basic understanding of the relations among the major groups of primates or of the time our species had branched off. The full story of human evolution still was not told; scientists continued to disagree on major issues, and something of an impasse developed.

The Evidence from Molecular Biology

Help then came from an unexpected quarter—molecular biology,[6] the study of the unseen, infinitesimally small structures and substances of all living bodies.

Figure 1.3 Bushman, adult male gorilla, as he appears in the Field Museum of Natural History in Chicago. Note how weight is borne by the thickened, hairless skin on the knuckles of the hands.

At the basic level of DNA (deoxyribonucleic acid, the genetic material of life) and the proteins that constitute the substance of most living things, biochemistry began to reveal the biochemical makeup of various species. Their likenesses and differences could then be assessed, because the record of evolution was mirrored in the chemical substances. Not all of it could be read immediately, or perhaps ever. Nevertheless it was there, and DNA was the determinant of the outer form that anthropology had been limited to studying in the past.[7]

The structure of DNA had been shown to be a helix (a kind of spiral), made up of two chains. In reproducing, the two chains separate, and each acts as a template for the assembly of a new chain. Biochemists were able to separate the chains and were soon separating the DNA chains of humans and many other animals.

In the formation of a new, hybrid chain—accomplished when DNA from one kind of primate is separated and mixed with the DNA from another kind of primate—only the correct matching units of one chain may attach themselves to the comparable units of the other chain. There must be a precision fit, and the fact that DNA units from a chimpanzee chain very closely fitted themselves to a human chain, with only a tiny degree of misfit, was startling evidence of the close relationship of the two species. The stability of this new DNA is dependent on temperature—the more closely related the animals, the more stable the hybrid. The similarity of human DNA to that of other primates is summarized as follows (the numbers measure the thermal stability in degrees Celsius):

Chimpanzee	*Gorilla*	*Orangutan*	*Old World Monkey*	*New World Monkey*
2.4	2.5	4.5	9	15

The comparisons are depicted in graphic form in Figure 1.4.

The relationships shown in Figure 1.4 are essentially the same Darwin believed them to be. Among all the primates, human beings are most closely related to the apes, particularly to the African apes. Study of

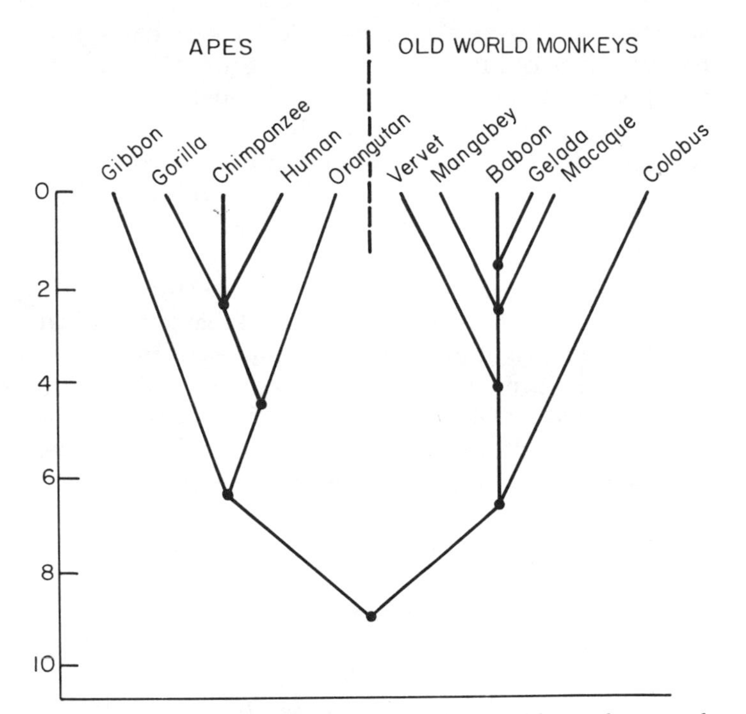

Figure 1.4 Relationships among Old World monkeys and apes as measured by comparisons of DNA. The comparisons are based on the thermal stability of the hybrid form—the greater the number, the less the stability. Numbers indicate degrees (Celsius).[8]

DNA offers no support to numerous other theories that have suggested that our closest living relatives are monkeys, or even tarsiers. What is surprising is the closeness of the relationship. We are as closely related to the chimpanzee as the chimpanzee is to the gorilla! Everyone had assumed that the two apes were far more closely related to each other than either was to human beings.

Conclusions based on the similarity in the DNA are quantitative, objective, and free from the kind of problems that result from anatomical comparisons. For example, after dissecting a great many primate hands, scientists might conclude that the human hand was basically that of an ape or that the human hand was primitive and had never passed through an apelike evolutionary stage. The similarities revealed by the comparisons of DNA are not open to this kind of subjective disagreement. Not only are human beings similar to apes, but the differences are remarkably small.

Another field of molecular biology brought in further evidence on evolution. This was a study of proteins. In the 1940s and 1950s Frederick Sanger worked out the sequence of the fifty-one amino acids in the two chains of the protein insulin. The order of amino acids is determined by DNA. Soon after Sanger's monumental work was done—and acknowledged by a Nobel Prize—many other proteins were similarly analyzed and mapped.

This research is fundamental in medicine and provides the background data necessary for synthesis of important molecules. New sequences are continually being published; so far, all support the following suggested relationships (the numbers represent the percentage differences in the chains of amino acids in the proteins that have been analyzed until now):[9] humans differ from chimpanzees (0.3), gorillas (0.6), orangutans (2.8), macaque (Old World) monkeys (3.9), capuchin (New World) monkeys (7.6). Human beings and the African apes are so close they can hardly be distinguished by this method.

Another science, immunology, supplied additional new data bearing on evolution. When a foreign substance enters the body, the body builds defenses against the intrusion. This immune reaction is of great importance in medicine and has been widely studied.

Shortly after the turn of the century, G. H. F. Nuttall began investigating what occurs when serum (blood minus the red cells) is injected into animals. They reacted by producing antibodies, and the antibodies were specific. When, for example, a rabbit was injected with human serum, it produced antihuman antibodies. In 1904 Nuttall suggested that immunology could be used to determine the relationships among various animals. He emphasized that the method would give quantitative results, not dependent on the judgment of the person making the comparison. Nuttall, however, was too far ahead of his time. Evolutionary studies were based on fossils and anatomy, and it was more than 50 years before the full significance of Nuttall's contributions was realized.

It also turned out that purified proteins were needed for accurate comparative studies. When they became available and the work was done, the chimpanzee reacted to antihuman serum almost as strongly as a human. The reaction was not as strong in the orangutan and was still less strong in the gibbon and monkey.

Extensive immunological surveys have been made by Morris Goodman at Wayne State University in Detroit and by Vincent Sarich at the University of California at Berkeley. Although they used different immunological methods, their results were essentially the same. Following are the results of Goodman's research[10] (the numbers are immunological distance units measuring reactions of other primates to humans in immunological tests; the smaller the number, the greater the reaction and the closer the relationship):

Chimpanzee	*Gorilla*	*Orangutan*	*Gibbon*
9	8	20	24

Macaque monkey	*Cebus (New World monkey)*	*Prosimians*
34	71	over 100

Figure 1.5 shows some of the differences between a macaque monkey and a prosimian. The immunological view of the primates according to Sarich and Cronin[11] is shown in Figure 1.6.

A fourth method of molecular biology—electrophoresis—furnished additional insights into the molecular relatedness of animals. Proteins differ in their elec-

Figure 1.5 The monkey, here a macaque, is active by day and has color vision. The brain is several times as large as the prosimian's. The sense of smell is less powerful. In short, the monkey is organized very much as we are. The lemur, pictured on the opposite page, is a prosimian. These contempo-

trical charges and may be separated by placing them in an appropriately arranged electrical field. Even small differences can be detected. Using electrophoretic methods, Bruce and Ayala showed that humans and apes are extremely closely related, as closely related as many species that have always been regarded as very similar.[12]

rary primates are very primitive, much like the earliest primates of 50 million years ago. They are nocturnal and small-brained, and the primitive sense of smell is well developed. Note that the nose looks more like that of a dog than that of a monkey.

Much of the new information, based on a variety of methods, has been summarized by King and Wilson.[13] They concluded that humans and chimpanzees share more than 99 percent of their genetic material and that they are as closely related genetically as sibling species, species that are regarded as being very closely related in comparison with the average differ-

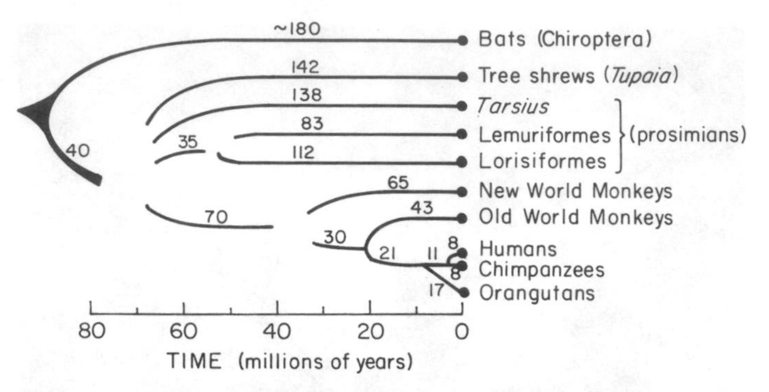

Figure 1.6 The numbers are immunological distance units that represent the reactions in the immunological tests. The smaller the number, the greater the reaction and the closer the relationship. Times are only approximate.

ence between species. Human beings and chimpanzees are even more closely related than horse and zebra, cape buffalo and water buffalo, cat and lion, or dog and fox. This high degree of similarity between humans and the African apes came as a surprise to everyone. On the basis of comparative anatomy, fossils, and observation, no one expected that the genetic differences were so small.

Each of these new molecular investigations thus says the same thing—humans are closely related to the African apes. The other primates are more distant relatives. Although the DNA differences are measured in degrees, the protein sequence differences in percentages, the immunological differences in distance units, all are in complete agreement. A summary is shown in Figure 1.7.

In later chapters we will try to show that these remarkable and surprising molecular findings are compatible with the fossil record and with evolutionary behavioral stages. Altogether, the recent work of science may end a hundred years of often acrimonious debate. As Darwin anticipated, it is turning out that the African apes are our closest living relatives. The monkeys, Old

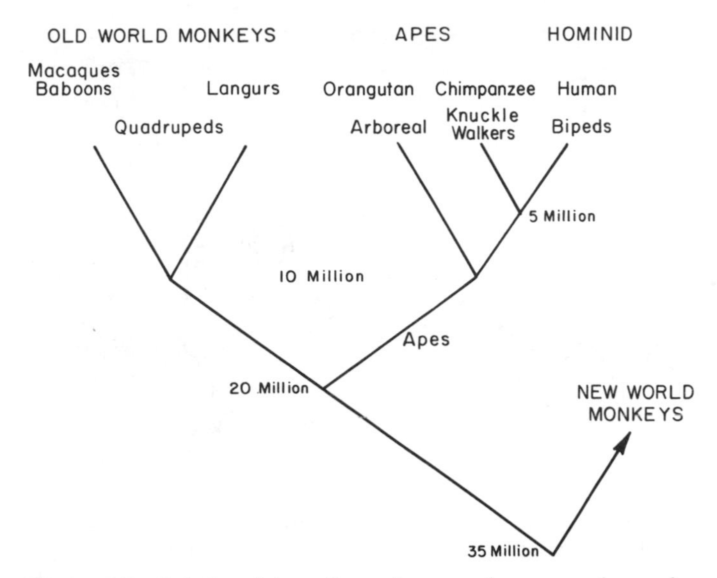

Figure 1.7 Relationships of monkeys and apes as shown by molecular biology. The scale: distance from human being to chimpanzee equals 1. Times are only approximate.

and New World, are more distant. The prosimians, once viewed as close ancestors of the human, are far removed. What is new is the method of proof and the elimination of many older but still surviving views to the contrary.

Time and the Fossil Record

As can be seen by looking at Figure 1.6, there seems to be general correlation between the immunological differences and time. Humans are recent, apes older, and monkeys older still. The prosimians are so different that they must have separated from other primates near the beginning of the age in which mammals have been dominant.

Attempts to time evolution by estimating when species separated often were met with incredulous opposition. The molecular-immunological data seemed

Figure 1.8 Time of origin of some major forms of life put on a 4,500-mile graph. See page 21 for fuller explanation.

to indicate that humans appeared very recently, perhaps 4 to 8 million years ago. Until recently few were willing to consider such a recent time span.

In 1800 the age of the earth was widely estimated at 100,000 years. A hundred years later the estimate went up to 20 to 40 million years. Strong efforts were later made to arrive at a scientific figure by estimating the depth of the sedimentary rocks and the accumulation of salt in the seas. All failed. In 1905 Ernest Rutherford suggested that the age of rocks might be determined by radioactive changes. But it was not until the mass spectrometer was invented and rapidly improved in the 1920s and 1930s that numbers of reliable dates became available.[14]

Dates were determined for earth (4.5 billion years old) and the beginning of life (3.5 billion years ago). Complicated cells with well-defined nuclei appeared 1.5 billion years ago; the first vertebrate fish, 500 million years ago.

Dates set in billions and millions of years are nearly incomprehensible to humans, who reckon a full lifetime at some eighty years. An analogy generally is required. Imagine, then, a graph with a scale of one mile for each million years. If it begins with the origin of the earth, 4.5 billion years ago, the baseline would be 4,500 miles long and would extend across the whole North American continent and out to the mid-Atlantic ridge. (See Figure 1.8 and Table 1.1.)

Some 1,500 miles westward, at about where Boston, Massachusetts, lies, life would begin. At Buffalo, New York, at the 2,500-mile mark on the graph, bacteria would have evolved. Another 1,000 miles west, at a point near Omaha, Nebraska, the first algae would be present; in another 1,000 miles fish would appear. Only at the last 70 miles would primates scurry onto the scene. Apes would not appear until the last 20 miles from the Pacific Coast, and human beings would show up only in the last 5 miles. On the scale of accurate radioactively determined time, humans are very recent—utter newcomers compared with fish, algae, or bacteria.

Agriculture, marking the beginning of modern technology, would be present only in the last 60 feet; and people like ourselves, *Homo sapiens*, would appear

Table 1.1 *Time since the beginning of earth.*

Origins of:	*Millions of Years Ago*
Earth	4,500
Life	3,500
Algae	1,500
Fish	500
Amphibians	350
Reptiles	300
Mammals	100
Primates	65
Monkeys	35
Apes	20
Humans	5
Homo erectus	1.5
Neanderthal	0.1
Homo sapiens	0.04

in the last 200 feet—about the width of a small strip of beach.

Unaided by science, the human mind is incapable of framing or comprehending the vast extent of time; certainly it failed to do so for centuries. Not until the last fractional inch of the graph's baseline did modern science establish the true age of the earth and discover that the surface of the earth itself is not immutable.

Continental Drift: Plate Tectonics

The discovery that the crust of the earth on which we live is a shifting, creating, destructing, erupting thing was another scientific revolution that affected the theory of evolution. If the continents were not fixed but moving, the movements of living things must have changed also. If seas expanded and contracted and mountainous barriers rose and eroded, routes of migration and contact were opened and barred. In all the changing environments some species survived and others perished; the course of evolution was directly affected.

As early as 1912 Alfred Wegener, a German meteorologist, suggested that the continents might have moved and that as great segments of them broke away from a solid land mass the oceans and seas filled in the gaps. Wegener could not explain how such masses could move, and his theory was largely disregarded.

Only in the 1950s and 1960s was the answer found: vast volcanic intrusions along midocean ridges were continually shoving the seafloors and continents away from the ridges. The motions of the continents have been very thoroughly described and the main events dated. The relation of the African plate to the Euroasiatic plate and the relations of Africa to South America are the points of importance in the understanding of primate evolution (see Figures 1.9 and

Figure 1.9 The world some 135 million years ago. The formation of the modern continents has started.[15]

1.10). The motions are based on complex geological processes which are clearly described in *Earth* by Press and Siever.[16]

Traditionally, it was thought that the monkeys of the New World and of the Old World were similar only because of parallel evolution—that both were descended from prosimians early in the age of mammals. It was believed that New World monkeys were descended from North American prosimians, but the molecular data show that both kinds of monkeys shared a very long period of common ancestry. Since South America and Africa were much closer together 35 to 40 million years ago than they have been in modern times, the New World monkeys' ancestors may have drifted from Africa to South America on some sort of natural raft; some rodents did likewise. Plate tec-

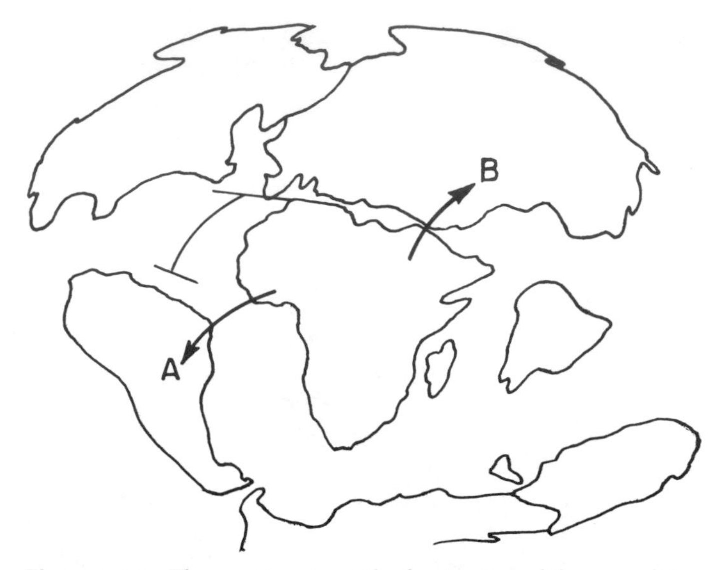

Figure 1.10 The continents at the beginning of the age of the mammals. *A* marks the possible route of the New World monkeys from Africa. *B* marks the route of possible migrations from Africa to Asia after Africa had drifted into Asia some 18 million years ago.[17]

tonics (continental drift), plus molecular biology, suggests a radically new view of much of primate evolution.

It is a new world. When Charles Darwin published *The Origin of Species* in 1859, there was no science of genetics or molecular biology, there were no methods for determining geologic time, and continental drift was unthinkable. There were no remains of early humans, and the nature of the Neanderthal skullcap was being debated. The evidence for human evolution came largely from the field of comparative anatomy and from the conviction that if all other forms of life had evolved, human beings must have been subject to the same laws of nature. Granting the limitations of the evidence, it is remarkable how right Darwin was! In publishing *The Descent of Man* in 1871, Darwin described how bipedal locomotion probably evolved before other specifically human characteristics. This thesis touched off a debate that was not finally settled until the discovery of human footprints at Laetoli, Tanzania, in 1978.

Over the last hundred years large numbers of human fossils have been found, and scientists have developed the methods necessary to settle many of the old problems. In the chapters that follow we will return to the implications of advancing technology and how it alters perspectives on human evolution. However, before discussing how the new world of modern science has clarified views on human evolution, one conclusion should be stressed: the difference between a modern view of human evolution and the theories of a hundred years ago lies in technology; the theories and facts of modern science are what provide the new perspectives. Human brains have not evolved in the last hundred years, and there is no reason to think that we are more intelligent than our ancestors of many thousands of years ago. In a very real sense, modern technology is managed by very primitive structures, human brains.

The study of evolution is one way of gaining some understanding of ourselves, and it is for this reason that the issues of human evolution have been so hotly debated. Modern technology will settle many of the issues of the times, places, and processes of human evolution; but the implications of all this progress will still be debated by the owners of emotional, primitive human brains. There will be no final answers to the problems of human evolution, or at least not for a long, long time.

2

HUXLEY WAS RIGHT

There was the great, seeming contradiction. Apes and humans obviously were different and no one could possibly confuse the two. Yet the newest findings were showing that down deep the two were remarkably alike. Such a combination, an apparent contradiction, was confounding and unexpected at this time. If the differences that were so obvious to the eye were only the finishing touches and the development of the last 4 to 8 million years, and if the likenesses were the heritage from an earlier ancestry, then there was no contradiction. Outer difference and inner likeness were explainable. Nevertheless, the problem was so complex and in some parts so new and controversial that all possible evidence had to be found. The urgency led to new studies of the anatomy of humans and apes.

Anatomy had been the first resource of those who set out to understand evolution. "It is notorious," Darwin wrote in *The Descent of Man,* "that man is constructed on the same general type or model as other mammals. All the bones of his skeleton can be compared with corresponding bones in a monkey, bat, or seal. So it is with his muscles, nerves, blood vessels, and internal viscera. The brain, the most important of all organs, follows the same law."[1]

Huxley, who made the bone-by-bone and muscle-by-muscle comparison, had concluded: "Whatever system of organs be studied, the comparisons of their modifications in the ape series leads to one and the same result—that the structural differences which separate Man from the Gorilla and the Chimpanzee are not so great as those which separate the Gorilla from the lower apes."[2] The similarity of general structure had been evident long before biochemistry disclosed the molecular likeness.

But the anatomical similarities did not effect agreement among the scientists. Highly trained individuals, using the same facts, judged our closest relatives among the contemporary primates to be as diverse as African apes, monkeys, or tarsiers. There was no concurrence on how anatomical facts should be interpreted or even counted.

Apes and Humans—The Differences

Although scientists believed in the theory of evolution by natural selection, little attempt was made to apply this theory to the interpretation of the differences between apes and humans. According to the theory, evolutionary success depended on behavior, and behavior had to be the scientists' guiding principle when making comparisons. By proceeding in this way, it was obvious that apes and humans differed in their way of moving on the ground and in structure of brain and face.

The human trunk proved similar to that of an ape in length of arm, breadth of trunk, and shortness of the lumbar region. More detailed examination showed that the similarity extends to the sternum; length of clavicle; and many details of the bones, joints, and muscles. In short, apes and humans share major structural features of the trunk and the motions that these make possible, such as stretching to the side and hanging comfortably by one arm.

This finding was dramatically illustrated when a scale drawing of a man was cut in half down the center, and a similarly scaled half drawing of an ape was placed beside it; the two trunks fitted together from thigh to shoulder almost as though they went together. If it had not been for the higher hunched shoulders of the ape, the fit of the body center would have been almost complete. (See Figure 2.1.)

It is in the whole middle section of the body that humans are essentially the same as apes. The structure and musculature make it possible for both to move their arms powerfully and easily to the side and above. For the ape these are the movements for climbing and feeding. To find out how essential the sideways and upward reach are, students sometimes are asked not to reach for anything for an hour. That means not picking up a book from a desk, or a plate in a cafeteria line, or holding on to a strap in a crowded bus. The ape reaches up to a branch while picking and eating some ripe fruit. It is the environment that differs for humans, not the way the arm is used.

Both apes and humans have a forearm capable of 180 degrees of rotation. Both thus can easily put the arm down flat on a table with the palm of the hand turned down or, with no conscious effort, turn the arm around with the palm upward in the begging or supplicating position. The structure that both share also permits strong flexion or turning in either position—that is, we can chin ourselves facing the bar with the palms of the hand turned either inward or outward. The

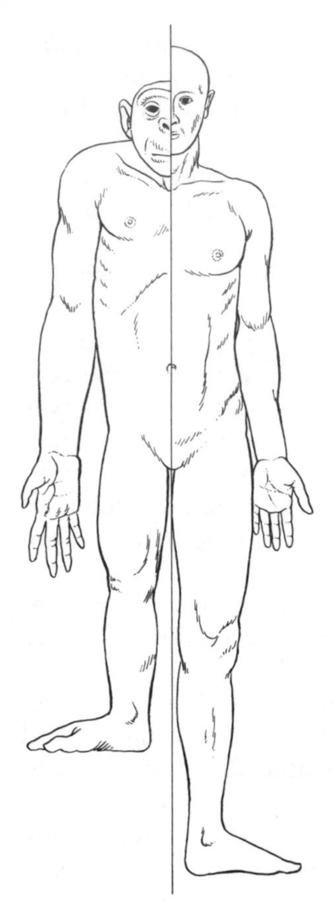

Figure 2.1 This half-ape, half-human drawing illustrates the outward physical similarities and differences between apes and humans.

monkey, in contrast to apes and humans, does not have the same degree of rotation or similar power. Apes and humans also have much freer movement of the wrist and elbow than do monkeys.

Monkeys also do not share an even more remarkable feature of the flexible ape arm—the stable position of the wrist in what the anatomist calls the "partially supinated position." This permits the chimpanzee, the gorilla, and the human football lineman to double under the ends of the fingers and rest part of their weight on their knuckles. The apes move very efficiently through the forest by knuckle walking. The lineman uses the knuckles-down position as he prepares to charge the opposition because it gives him a fast getaway (see Figure 2.2).

Monkeys have a different wrist joint; when they move along the ground, their hands are put down flat. They are better adapted to moving in the trees than on the ground. The monkey palm also is long; it covers the lower end of the ulna, or forearm bone. In apes and humans this elongated palm has been lost.

A seeming oddity further underscores the likeness of humans and apes. Neither has much hair on the backs of the fingers, and this relative baldness is under genetic control. Scientists formerly could not see any advantage in relative hairlessness of the finger backs, but it is this surface that bears the weight of the knuckle walker. In the chimpanzee and the gorilla the skin over the middle bone of the fingers is thick and hairless, a common development of any weightbearing surface.

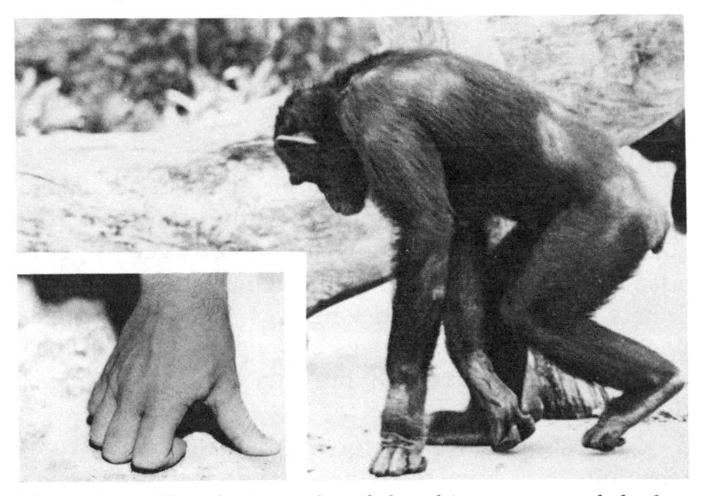

Figure 2.2 The photographs of the chimpanzee and the human hand (inset) show the similarities between the chimpanzee knucklewalking and the human hand in knuckle-walking position.

It may well be that this lack of mid-digital hair in some people is at least a partial adaptation to knuckle walking by our remote ancestors.

In summary, the structure of the human trunk and arms is remarkably apelike. It is this anatomical similarity that forms the basis for the actions that are similar in apes and humans. Just as the motions in the living creatures are basically alike, so the structures are alike. The fact that anatomists have no difficulty in using an atlas of human anatomy when dissecting a chimpanzee arm clearly shows this.[3]

For example, in the short, wide, shallow chests of apes and humans, the heart and lungs are located in the same position, differing greatly from their positions in the long, narrow, deep chests of monkeys. In apes and humans the colon has ascending, transverse, and descending portions, whereas in monkeys the colon has no comparable divisions.

A second great complex is made up of the legs, where the differences between humans and apes are striking. In the half-and-half comparison, illustrated on page 30, the human leg is ten or twelve inches longer than that of its ape counterpart. A composite creature would walk with an impossible limp. Upon examination, bones, muscles, and functions of the lower legs of the two are easily distinguishable.

Neither an ape nor any other animal walks like a human, although there are many other bipeds. Birds, some dinosaurs, and pterodactyls evolved bipedal adaptations without freeing the hands for tool using. Others also have evolved long legs. Tarsiers and gibbons, for example, have legs as long as humans' relative to the trunk, but only humans walk like humans. (See Figure 2.3.)

The structural basis for the human walk is complex; initially it involved a new method of transferring the landing and balancing functions of the forelimbs to the hind limbs. In the quadrupedal monkey or knuckle-walking chimpanzee, when the drive for a step comes

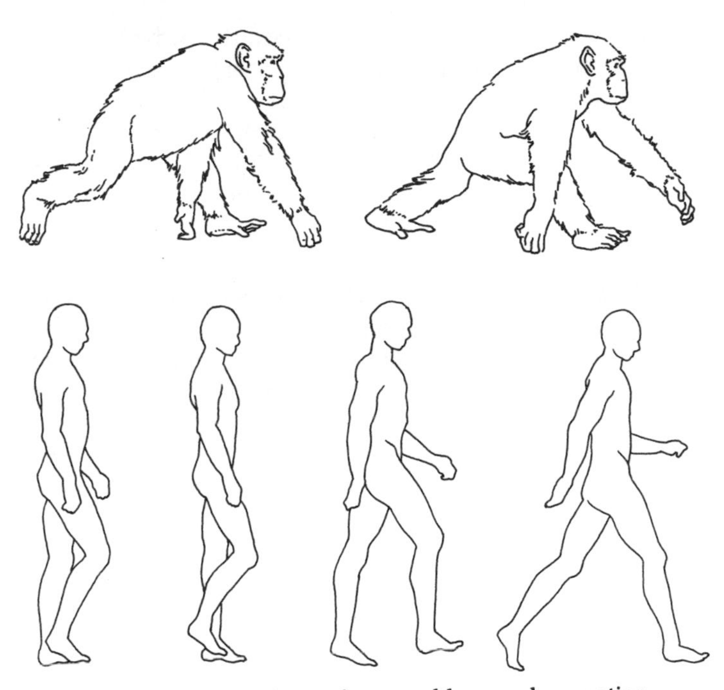

Figure 2.3 A comparison of ape and human locomotion.

from the right rear leg, the animal lands on the left forelimb. The human foot has two functions to perform as it touches the ground: it must first land and balance, and then it must give a push to produce the next step.

To make these functions possible, and to maintain the vertical position, the pelvic region had to change. The ilium, the upper part of the pelvis, became shortened in comparison with that of the apes. With related changes in the sacrum, the bottom section of the vertebral column, a bony birth canal was completed. However, the shortening of the ilium alone did not change the relation of the trunk and legs and eliminate the stoop of the apes. For the backbone and trunk to become vertical a backward bend of the ilium also was necessary. But without further changes the backward bend would have blocked the pelvic outlet, and an animal born with such a change could not have given birth to offspring. But other changes did occur; the sacrum moved up and out of the way of the birth canal, and a curve was added to the lower backbone, making the trunk vertical. There were some lesser changes also, and the result was that the knuckle-walking ape became upright and could walk in a new way (see Figure 2.4).

The discoveries at Laetoli discussed in Chapter 1 indicate that human bipedal locomotion evolved 3.6 million years ago. As mentioned, a remarkable find was made in this locale in the summer of 1978 by Mary Leakey and her co-workers—human footprints. Most early fossils are fragmentary and limited to teeth and jaws, and until recently there were no radiometric

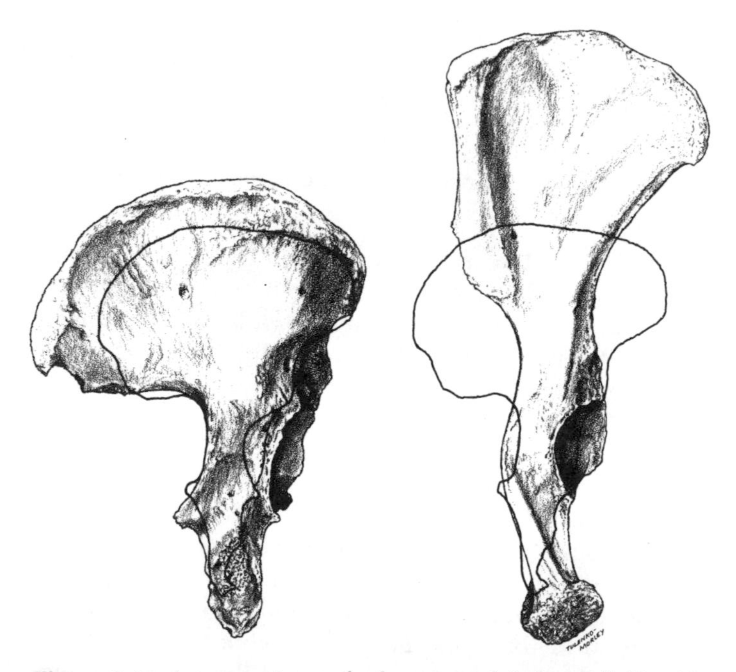

Figure 2.4 A comparison of a human pelvic bone (left) with that of a chimpanzee (right). The outline of the same bone of *Australopithecus* has been superimposed on the others.

dates. The interpretation of the limb bones was debated for some forty years, and although many scientists thought the fossil evidence proved that bipedal locomotion evolved long before large brains and some other human characteristics, they could not prove this. The Laetoli footsteps substantiate that bipedalism did evolve a minimum of 2 million years before large brains developed.

The next big difference in the half-and-half illustration is in the face. The ape's big muzzle, massive jaw, and heavy brow ridges seem far from the human face. An ape fights with its face; its canine teeth are sharp and strong enough to rip and tear an enemy and are powered by strong jaw and neck muscles. Only when tools supplanted dentition as weapons could there be a change in selective pressures for big teeth and all that went with them.

The reduction of the canine complex was an evolutionarily novel characteristic of the Hominidae. In terms of behavior it meant that tools took over the functions of the teeth in protecting against predators, in fighting within the group, and in fighting between groups. Our ancestors thus lost their "frightful physiognomy" and took on the face that we regard as less brutal and more refined. (See Figure 2.5.) Recent field studies show that male apes fight considerably more than was believed to be the case even a few years ago. We will return to this critically adaptive behavior in more detail.

As Darwin pointed out, reduction in the canine complex occurs primarily in males. Female gorillas have small canine teeth, as shown in Figure 1.1 on page 4. If we can learn why the male canines are large and female canines small in many kinds of monkeys and apes, we can then understand why this kind of sexual differentiation was lost in human evolution.

Another anatomical difference separates apes from humans. In the half-and-half illustration (Figure 2.1), the human's head rises much higher than the

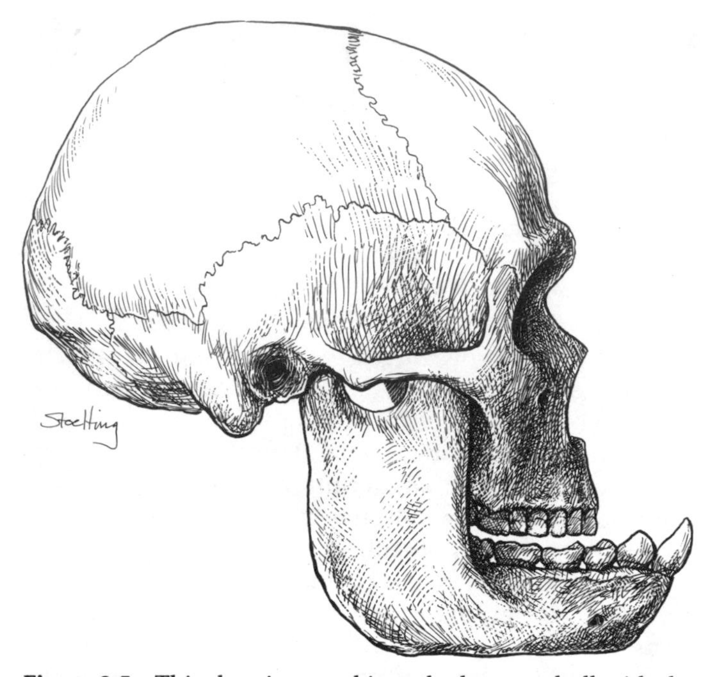

Figure 2.5 This drawing combines the human skull with the jaws of an ape. Notice that the lower jaws of the ape project far beyond the upper teeth of the human skull; the teeth contrast as well, the ape's being larger and longer.

ape's low flat skull. At the top, the two halves do not match at all, the reason of course being that the human skull has been enlarged to house a larger brain.

The average human brain is about triple the size of the brain of the ape—some 1,200 to 1,500 cc as compared with 400 to 600 cc for the apes. The human brain resembles the basic ape brain but with several new added areas, which make possible hand skills, speech, improved memory, and the use of memory that involves conscious thought and planning. (See Figure 2.6.)

The areas can be mapped almost as clearly as the geologist maps a continent, identifying each area and

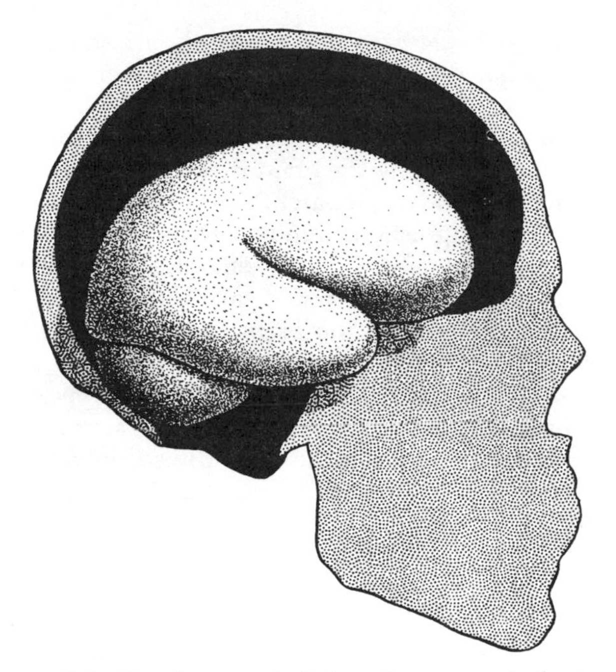

Figure 2.6 The human skull housing an ape's brain; the black area indicates the space the human brain fills.

its extent. Surgeons as well as other scientists helped in this task. As they operated to remove growths or to repair other damage, surgeons learned that certain areas produced certain responses. Gradually the human brain as a whole was mapped. The brains of apes have similarly been marked out.

A large part of the motor cortex of the human brain governs hand movements. In monkeys the hand and foot areas of the brain are about equal in size, whereas in humans the hand area greatly exceeds the foot area.

Part of the hand movements of both humans and apes are coordinated in the cerebellum, an area lying behind and below the brain, or cerebrum (see Figure 6.7 on page 169). The cerebellum in humans is approximately three times that of apes, and the most enlarged part is associated with learned hand movements.

Humans also have a large area of the brain used for language. The ape has only a small area for communication. Obviously the brain is essential for any act of communication, but apparently very little brain is required in nonlinguistic communications systems. Any debate over the nature of the neural structures that make language possible should take into account that there are new parts of the brain that make language possible. It is important to realize that apes and monkeys cannot be taught to talk, although they can be conditioned to respond appropriately to sounds and even to manual signs. In the low-browed, low-domed skull of the ape is a brain well provided with primary sensory and motor areas.

Behavior in the Wilds

Questions nevertheless persisted. Are the apes of Africa today, the gorilla and the chimpanzee, like any ancestral ape? Could studying the living apes supply any clues to the ancestral lines? Could a study of living apes tell humans anything about themselves?

Darwin and all evolutionists have emphasized that the living apes have evolved since the time—whenever it might be—when ape and human ancestors parted company. All conceded that change would have occurred at that time. New studies show that human behavior was evolving and changing rapidly and that apes stayed generally the same.

Humans were very slow to learn about their animal relatives. The first accounts—generally sailors' tales—of the existence of humanlike apes began to reach Europe in the seventeenth century. In *Purchas His Pilgrimage,* published in England in 1613, an old soldier who had lived in Africa for many years reported "a kinde of Great Apes, if they might bee so termed, of the height of a man, but twice as bigge in features of their limmes, with strength proportional, hairie all over, otherwise altogether like men and women in their whole bodily shape."[4]

Drawings of the reputed "monsters" often resembled a human in a hairy suit; a tail was sometimes added (see Figure 2.7). In 1699, though, the Royal Society published a memoir listing forty-seven points in which the "ourang-outang" "more resembled Man than Apes and Monkeys do," and thirty-four points in which it "differ'd from a Man and resembled more the Ape and Monkey kind."

Few were willing to undertake any kind of a systematic study in the field. As Huxley said in 1863, "the man who risks his life by even a short visit to the malarious shores of those regions [Africa and Asia] may be excused if he shrinks from facing the dangers of the interior . . . and contents himself with collecting and collating the more or less mythical reports . . . of natives."[5] If the dangers of the interior did not discourage possible students of the interesting great apes, reports of their ferocity certainly did. When one investigator went to Africa in 1896 to study the gorilla, he built him-

Figure 2.7 An early drawing of a primate.

self an iron cage in the forest and sat there day after day waiting for the gorillas to appear. Needless to say, they did not.

Field Studies: The Chimpanzee

A few studies got under way before World War II, but the rush to study apes in their habitat started in the 1950s. Investigators from Japan, England, France, Switzerland, and the United States launched field studies of the behavior of a variety of primates. At last some of the answers to evolution were to be sought in animals living as humans' ancestors must have lived in the forests of the past.

By the end of the 1960s more than twenty studies were reported, most based on at least 1,000 hours of close observation, and some, like Jane Goodall's, on continuing years of work.[6]

Goodall, with an early and continuing interest in animals, left school at eighteen to work until she could go to Africa to study animals in the wild. Soon after she arrived in Nairobi, Kenya, in 1957, she met Louis Leakey. When she told him of her interest, he suggested that she start by working for the Coryndon Museum of Natural History in Nairobi. For a while she served as Leakey's secretary. Later she accompanied Mary and Louis Leakey to their dig at Olduvai Gorge, Tanzania. Day after day they worked, picking at the rock face with dental probes. The Leakeys' Dalmatians stood guard, for rhinos often wandered onto the site, and at night lions could be seen around the camp. In the evenings, nevertheless, Goodall explored the country to watch the giraffes and the lions.

When Leakey became convinced that her interest in animals was not a passing fancy, he suggested that Goodall study the behavior of chimpanzees and assisted her in getting a grant from the Wilkie Foundation to start the work. She arrived at the Gombe Stream Reserve in June 1960 and immediately began her search. But the chimpanzees fled the moment she approached. She could only hear them calling from afar. "I often returned to the camp in utter discouragement," she said. "Was my whole attempt doomed to failure?"[7]

Gradually the chimpanzees began to accept her. They no longer fled when they caught sight of her, and she began to make notes on their feeding behavior, the changing composition of the groups, and their habits of nesting in the trees each night.

In time she could sit close enough to watch the chimpanzees without binoculars. They became individuals to her—Flo with the torn ears and her infant daughter Fifi and her adolescent sons Figan and Faben; David Graybeard with his white beard and gentle disposition; Leakey; J.B., for John Bull; and many others. (See Figure 2.8.) In about four months Goodall knew the fifteen miles of rugged country the chimpanzees frequented.

One morning she came upon David Graybeard trimming the edges from a wide blade of sword grass. He then poked his cleaned rod down a hole he had scratched in a large, domed termite nest, waited a few minutes, and then skillfully pulled out his rod and delicately licked it with his lips. He was termite fishing. When his first fishing rod weakened, he picked a piece of vine, stripped off its leaves, and continued his successful pursuit. Every now and then he would put his ear to the nest, listen carefully, and then scratch a new fishing hole. Only in certain seasons do the termites make tunnels out to the surface. Until the time comes for a nuptial flight away from the nest, they keep the surface sealed with a thin cover of clay. When the chimps discovered and opened the tunnels, the termites resealed them as quickly as possible. (See Figure 2.9.)

Goodall realized that some animals could use objects as tools and that a chimpanzee had been observed actually making a tool. In her first year of work Goodall

Figure 2.8 Faben, Figan, and Fifi play around Flo and Flint. These chimpanzees and others, whose photographs appear in the rest of this book, were observed by Jane Goodall and photographed by Hugo van Lawick.

had made a finding of major importance. In the past the human being often had been defined as the "toolmaker." It was the supposed critical difference that set us apart from all other animals. As Goodall's and others' work confirmed, the definition would have to be redrafted. Humans were redefined as beings who

Figure 2.9 Jane Goodall observed chimpanzees using tools —grass blades, twigs, and vines—to catch termites, a favorite food.

can make tools according to a set, predetermined pattern.

Two years later, while sitting on the veranda of her tent, Goodall saw one of the large male chimpanzees enter the clearing. At first, with complete composure, he climbed one of the palm trees and ate some of the ripe nuts. Then he came down and, with hair bristling, approached a table and took a banana lying on the table. Goodall had her cue—after that she began supplying bananas, and many of the chimpanzees came daily for this fare. To prevent the big males from devouring everything (one large male has been observed eating sixty bananas at a sitting), Goodall began distributing the fruit around the grounds in separate boxes. The chimpanzees either tore the boxes open or took them apart by pulling pegs and opening catches. Goodall finally had to devise concrete boxes with steel lids, controlled by wires threaded through underground pipes. When the animals mastered this system, Goodall finally had to turn to electrically operated boxes.

Field Studies: The Gorilla

George Schaller was a graduate student in zoology at the University of Wisconsin when one of his professors, John T. Emlen, asked if he would like to study gorillas. Two years later Schaller was in Africa, in the Congo. As he and a guide followed a trail of broken branches and tramped grass left by a group of gorillas, his guide muttered constantly, "Gorilla kill you." Suddenly Schaller heard a rapid "Pok-pok-pok." A gorilla was beating its chest. Schaller had encountered his first gorilla.

In a short distance they came within sight of an adult male, sitting among the shrubs and vines. He was easily identifiable by his huge size and his silver gray back, and beside him sat a juvenile and three females, "fat and placid with sagging breasts and long nipples." Up in the fork of a tree was a female with a small infant clinging to the hair on her shoulders. "I was little prepared for the beauty of the beasts before me," said Schaller in his book, *The Year of the Gorilla*.[8] "Their hair was not merely black, but a shining blue black, and the faces shone as if polished." Schaller and the gorillas sat watching each other. The big male rose repeatedly to his full height of about six feet to beat a rapid tattoo on his bare chest and then sat down again. After a time the group walked quietly away. There was none of the fabled ferocity of which Schaller had been warned, and later the gorillas came to accept him to some extent. The fierceness proved to be a myth. Schaller admitted that his hair never failed to rise when he heard the shattering roar of a big silver-backed gorilla, but he soon learned that the bloodcurdling sound and the huge rearing body were largely bluff. The hurtling bulk nearly always stopped before making contact with an opponent. "Gorillas are eminently gentle and amiable creatures," Schaller wrote. "Peaceful coexistence is their way of life."

Schaller's observations also disclosed that gorillas have strong attachments to members of their own groups. When groups met as they ranged through the forest on their daily rounds, each tended to stay together but did not fight over territory: "The gorilla certainly shares its range and its abundant food resources with others of its kind, disdaining all claims to a plot of land of its own."

Schaller found that the gorillas spent from 80 to 90 percent of their waking hours on the ground; they were not primarily arboreal, as most accounts had indicated. When they did climb trees, generally to sun themselves and sometimes to make a nest, their actions were slow and deliberate. Schaller compared their climbing ability with that of a ten-year-old child.

As the gorillas moved along the forest floor, they walked with their knuckles down, often in a procession led by a silver-backed male, and brought up at the rear by a black-backed, younger male. The females and young generally were between them. Often the troop traveled from 1,000 to 2,100 feet a day in their ten-to-fifteen-mile range. They did not return at night to a home base. "In some respects, gorillas might be considered nomads," Schaller wrote, "moving along and feeding, and finally bedding down whenever darkness overtakes them, only to begin another day of wandering."

We will return to the behavior of the contemporary primates on page 169 in Chapter 6. Here we are concerned only with the major behaviors necessary for a background for understanding human evolution, and Goodall's and Schaller's work is stressed because of the importance of the chimpanzee and gorilla in this respect. There have been numerous other important field studies recently, especially Dian Fossey's work on the gorilla and Birute Galdikas's on orangutans. Japanese scientists, particularly Junichiro Itani and Toshisada Nishida, have made many contributions to the field studies of monkeys and apes as well.

The Past Way of Life

The field studies built up information on many primates, on their characteristics and their individual and social behavior. Such information was, as Phyllis Jay of the University of California and editor of *Primates —Studies in Adaptation and Variability*, noted, "a glimpse into what may have occurred as man evolved into man."

> A great deal can be learned from the bones that comprise our fossil records, but the life of ancient primates comprised much more than the obvious function of these bony parts. For example, a certain kind of roughened surface on a fossilized ischium [the seat] merely indicates that the primate has ischial callosities. But by looking at how living primates with these callosities behave it is possible to infer that in all likelihood the ancient animal slept sitting up rather than on its side in a nest. This may seem trivial, but when many such clues are gathered and collated, the total picture of an animal's way of life fills in to a closer approximation of what it must have been.
>
> Our ancestors were not the same as the living primates, but the rich variability of behavior of modern monkeys and apes makes it possible to reconstruct the most probable pattern of related forms in the past.[9]

The studies of animals in their habitats are making these suggestions about what life was probably like for our ancestors:

1. A few years ago it was thought that the social behavior of the great apes was similar. Now it is clear that the fundamental adaptive behaviors are different. For example, the orangutan is solitary. The only social group consists of a female and her young, although occasionally a male may join the group in order to mate. In chimpanzees there is a moderately large social group, but it often divides. In gorillas the group is much more compact. We will return to the reasons for these differences on page 138; but the solitary orangutan is clearly the exceptional species. Most monkeys and apes are intensely social.
2. The great apes occupy limited areas, and these may be defended. Their diet is primarily vegetarian, and the group may support itself in a few square miles. Local population densities may be very high. When our ancestors became hunters they required far larger territories, and this led to lower local densities. In this way successful hunting changed the whole basic genetic-demographic pattern for our species. The evidence for hunting in the fossil record is critical for the reconstruction of the behaviors of our ancestors.
3. When we wrote the first edition of this book, we thought it would be possible to make at least an informed guess about the sexual behaviors of our ancestors, but recent studies show that the sexual behaviors supposed to be major differences between ape and human do not exist.
4. Under natural conditions, the nonhuman primates rarely manipulate objects. Exceptions are the great apes, particularly the chimpanzee. All the apes throw sticks toward enemies, but chimpanzees

throw stones as well; and they use leaves, twigs, and branches for a wide variety of purposes. The great apes make nests, which involves elaborate, stereotyped motor patterns. It is clear that the contemporary apes are far closer to human beings in their manipulative ability than is any monkey. We think that the anatomical basis for these abilities is in the joints and might be detected in the distant past if more complete fossils are found. For example, apes and human beings have great freedom of movement in the shoulder. This is clearly reflected in the shoulder joint. The upper end of the armbone (the humerus) is remarkably similar in ape and human and very different in quadrupedal monkeys. If a fossil of the humerus could be found, the kind of movement it was capable of might be analyzed. Unfortunately, the bones that would provide a better idea of the evolution of the shoulder have not yet been found.

Just as molecular biology, dating methods, and continental drift form a new framework for viewing human evolution, so the field studies of behavior provide a new basis for considering the evolution of behaviors. Ape behavior is by far the closest to human behavior, but it is much more variable than people suspected. We learn both from similarity (the importance of play) and from difference (the lack of practice for manual skill). Some knowledge of behavior is essential for reconstructing the past, as the reader will repeatedly note in the pages that follow.

"In summary, the outstanding difference between human and nonhuman use of tolls is *skill* and the biology that makes skill possible," said Jay. "Many primates use tools, but only a few species, apes and men, use tools as objects in agonistic display."[10]

All observers also have pointed out the importance of play among young nonhuman primates. It is how they learn the behavior that will be required of them as adults. The play behavior of chimpanzees and gorillas is remarkably more complex than that of monkeys. "Surely it is more than a coincidence that [the chimpanzee] the nonhuman primate taxonomically closest to man is, according to many investigators, also the most manipulative, exploratory, and similar to man in play," Jay wrote. "The range of variation in play forms and games among chimpanzees is second only to man."[11]

The young apes spend much of their time playing, running, and tumbling; clambering over adults; swishing branches and vines; and even rolling and using hard round fruit as balls. Nearly all monkey play, on the other hand, is wrestling and chasing; there is relatively little play with objects.

The new studies demonstrate that the social and individual habits of apes are much more like humans' habits than those of monkeys. Much of the base of human behavior is evident in these studies of apes in their natural environment. Much of it carried over. Where humans depart from the behavior of the living primates and from the probable behavior of our ancestors, as in skilled use of tools and wider range, the departure came late. The animal studies, the days and years spent quietly in the forests watching the animals, added confirmation to the other work indicating that the development of humans came in the last 5 to 10 million years.

From Ape to Australopithecine

One species did not stop and another start—it was not like that at all. Over several millions of years, some early primate populations that may have ranged over much of Africa and Asia developed a few human or humanlike characteristics. Before the australopithecines, however, the fossil record is too incomplete to permit many certainties. No one can say positively which of the changing apes were in the direct line to the australopithecines. Fossil apes with varying characteristics have been found in Europe, Africa, and Asia.

Far back in time, in the distant reaches that are nearly incomprehensible, forests covered most of Africa, Asia, and Europe. The continents were joined at many points, and there were no geographic barriers to the movement of apes. One genus might extend from Africa to India, a range indeed not greater than that occupied today by the African vervet monkey *Cercopithecus aethiops*. And one or more did.

The fossilized teeth and a few other scraps of the bones of prehuman primates have been recovered from sites in Spain, France, Kenya, Uganda, Turkey, Hungary, the U.S.S.R., Egypt, India, Pakistan, and China. Elwyn L. Simons of Duke University estimates that the number of specimens may now total many hundreds.

Some of the Finds

About sixty miles southwest of Cairo where a desert wasteland now presses close to the brackish Lake Qarun, meandering tropical rivers once flowed into the sea. Many animals lived in the forests along their banks, and fossil hunters have been finding their remains for more than a half century. When Yale expeditions led by Simons went there in the 1960s to search for fossil primates, they found they often had only to remove the overlying rock, or better yet, sweep away the "desert pavement"—rock cover—and permit the wind to scour out tons of unconsolidated sediments.

In one fossil wood zone they discovered a most unusual skull. It looked like the skull of a monkey, though the teeth were much more like a gorilla's. The canines were large and the front premolars elongated. Simons accurately described the creature as a monkey with the teeth of an ape and named it *Aegyptopithecus* (see Figure 2.10). In the same area, wind action incovered several tailbones that seemed assignable to *Aegypto-*

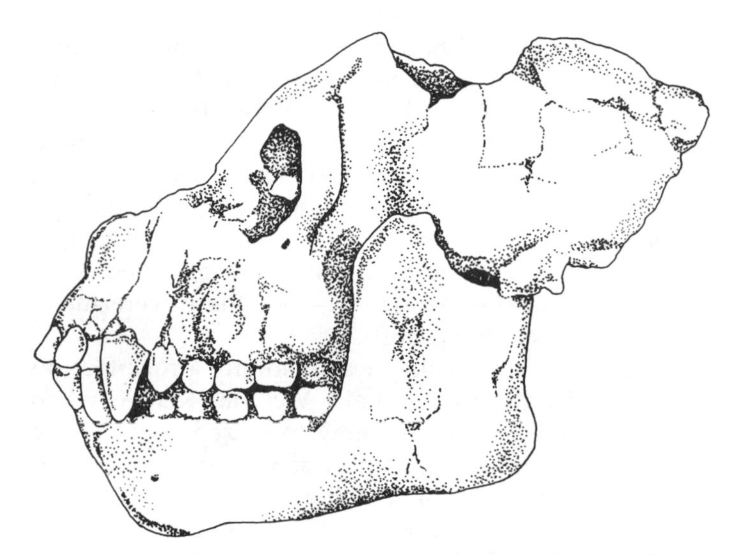

Figure 2.10 The partially restored skull of *Aegyptopithecus zeuxis,* so named by Elwyn L. Simons, found in the Fayum region of Egypt. The lower jaw is a restoration based on jaw fragments not found in association with the cranium; the incisor teeth of the upper jaw are also restorations. The skull belongs to a species of ape probably ancestral to the dryopithecine apes (lit., forest apes), and dates back between 26 and 28 million years.

pithecus, which did not surprise Simons: "Speaking anatomically, it was to be expected that some ancestral primate would cross the threshold separating monkey from ape and still bring its tail along."

The eyes of *Aegyptopithecus* had shifted forward, giving it the better depth perception that would be invaluable in the trees. The creature was big for a tree dweller, and its feet—some foot bones were found—also suggested adaptation to life in the trees. *Aegyptopithecus* bore the stamp of what was to come. "The animal was evidently pursuing an arboreal pattern of life directing it along the evolutionary path leading from lemur-like and monkey-like forms to apes and perhaps ultimately to man," said Simons.[12]

A step onward had been taken. For nearly a century investigators had been digging up teeth and jaw fragments of a fossil primate that also did not fit into the pattern of any known apes. Simons commented that if *Aegyptopithecus* could be called a monkey with the teeth of an ape, these primates might be dubbed apes with the bodies of monkeys. The head was apelike. As a group, the odd assemblage was called dryopithecine—literally, forest ape—but there were many variations within the group.

In 1948 Mary Leakey was scanning a rock face on the island of Rusinga in Lake Victoria, Kenya, when her eye caught a speck of gray fossilized enamel. She and her husband, Louis, who was working nearby, immediately began to dig out her find. Behind the exposed tooth was another tooth, and behind that was something more. In several days of painstaking work they dug out a nearly complete skull. It had a rounded forehead like that of a human, but it also had long, pointed canines and other apelike features. The Leakeys named their discovery *Proconsul africanus,* and it was later dated at about 24 million years. "We believe," said Leakey, "that at some stage just about the time of *Proconsul* the stock that ultimately led to man broke away from *Proconsul* himself or from something much like him, and gradually led to you and me. Mary's discovery gave science the first opportunity to see what *Proconsul* had really looked like. Previously we had only jaws and teeth to go on."[13]

Often each discoverer of a major fossil, or even of a tooth that seems to differ from other teeth, tends to set up a special species and sometimes even a genus for its onetime owner. The imposing names given each fossil or its kind have multiplied. Others who have studied the fossils like to lump many of them together.

Simons, an authority on primate fossils, calls both *Proconsul* and the Siwalik finds dryopithecines but divides them into two series, a larger and a smaller. The larger form, including *Proconsul,* generally has a large snout, protruding incisors, and rather high-crowned teeth. He said: "May it not be that these two sets [*Proconsul* and Siwalik] represent a single species that ranged fairly widely and perhaps over a long period? . . . This species could well be ancestral to the gorilla and chimpanzee."[14] Simons suggested that the second and smaller Indian form might have given rise to the orangutan.

In the early 1960s, about fifteen years after finding *Proconsul,* the Leakeys were working at Fort Ternan, a fossil-rich site in Kenya. When Leakey returned after a short absence, his chief African assistant, Heslon Mukiri, had something special waiting for him. He lifted the lid of a box, and Leakey saw a jaw that he knew at once was important. It had what anatomists call a "canine fossa," a depression that occurs in the upper jaw of *Homo* just below the eye socket. It is never seen in the same form among fossil or living apes and monkeys. In humans it is an anchor for a muscle controlling the movement of the upper lip, particularly in speech.

The presence of the fossa did not indicate that this creature just unearthed from the past had the ability to speak, only that the potential for speech was developing. The canine teeth also were small and more like those of a human than an ape. The face was short. Some years later when a lower jaw was discovered in the same area, it was less human in conformation.

At the time of the first find, Leakey named the group represented by the jaw *Kenyapithecus wickeri* (Kenya ape, and *wickeri* for Fred D. P. Wicker, who had financially assisted his work). Jack Evernden and Garniss Curtis of the University of California dated *Kenyapithecus* at about 14 million years. Similar jaws and teeth previously had been found in the fertile Siwalik Hills. They also had similar fossae and the U-shaped dental arch typical of humans rather than the V-shaped arch of the apes. The face generally was short, not at all an ape's snout. This peculiar ape had been named *Ramapithecus*.

Simons proposed that *Ramapithecus* and *Kenyapithecus* were much the same kind of creature, and perhaps just advanced dryopithecines. Because the forests of Africa and Asia were continuous at the time, he argued that this was not at all impossible or improbable:

> Separately almost all of these features [of the ramapithecines] can be found among the pongids [apes] but their occurrence in combination in *Ramapithecus brevirostris* is a strong indication of hominid ties.
>
> The transitional nature of these specimens of itself raises the question of the arbitrariness of separating the families Pongidae and Hominidae—a problem which has also been posed recently in connection with another event, the discovery of close biochemical similarities between man and the apes, and in particular the African apes.
>
> Personally I do not see that it very much matters whether members of this genus be regarded as advanced pongids or as primitive hominids.[15]

Advanced apes or primitive humans? The gap or reputed gap between the two was being closed. From ape to human, Simons was proposing, there seemingly was no break, only a gradual becoming, a continuum, a shading from one to the other. It was a transition that might have taken place over millions of years and over a vast part of the earth.

Answers and New Questions

The full story was not told. But much of it was there. The newer studies of anatomy and of the living apes, along with the restudy of the fossils recovered from the earth, all indicated that some early apes slowly and gradually evolved into the part human, part apelike australopithecines that launched the human line. Humans were meeting their remote ancestors—the apes, modified apes, and largely human apes.

3

TOOLS MAKYTH MAN

After the African continental plate drifted into the Eurasian plate some 17 to 18 million years ago, there were broad connections between the two great land masses. Numerous animals, including monkeys and apes, could migrate freely from one to the other, and it is more useful to think of one great undivided area than to think of the continents as they are today. We should think of *Ramapithecus* as having been in the center of this ancient world rather than having been in Asia, Europe, and Africa.

For some millions of years after the supercontinent was formed, the climate was warm. Woodlands covered much of the area, and probably apes were most numerous at that time. Later, some 5.5 million years ago, deserts formed and the Mediterranean basin was flooded by the Atlantic Ocean. More recently it became much colder, and the northern areas were repeatedly covered by ice. The warm, undivided Old World that had lasted for many millions of years was gradually transformed into the world we know today. Climatic changes during the last million years were particularly great. For example, Arabia was well watered much of the time, and when the ice advanced all the great islands of Southeast Asia (Sumatra, Java, and Borneo) were connected.

Even with the altered climate and reduced area there was plenty of room for the primates, and they must have been numerous. If a density of ten per square mile is assumed, and that is low, there would have been 20 million animals in the transitional populations of ancestral apes at any one time. Naturally, calculations of this kind may be far from representing the actual facts, but they are introduced to correct the impression that human origins occurred necessarily in one restricted small place in one short period of time. Monkeys were far more numerous than apes. For both monkeys and apes the canopy of the thick green upper world of the forest was a relatively safe place. A leopard might climb into the heavy lower branches of a thorn tree, but no carnivore could follow a monkey or even the heavier apes out to the ends of the branches. Even a 150-pound ape could safely move far out on a slender limb.

The top of the forest also was a world of plenty. Fruit and tender buds grew on the branches, usually toward their ends. But as secure and rich as the arboreal world might have been, it also had its deficiencies. Fruits and buds and insects at times became scarce, perhaps because the more agile monkeys may have beaten the apes to them. And perhaps at times the competition of the apes made life difficult for the monkeys, since the bigger apes sometimes drove them away from the best fruit.

But other fruit, insects, and roots could be found on the ground. An ape or monkey that came down from the trees also might be able to pick the fruit ripening in a grove of trees too isolated to be reached by leaping from tree to tree. Lions, hyenas, and other carnivores were a greater menace on the ground, but if trees were nearby an animal on the forest floor could quickly escape by climbing. The ground offered many advantages, despite its dangers. For some, or possibly all, of these reasons, and perhaps for other reasons, some monkeys and apes came to the ground. Although an unknown number of monkeys descended to the ground, most monkeys could not survive there; they perished and their species became extinct. Some of their fossilized bones have been found. Only three groups—the patas, the baboons, and the geladas—succeeded as primarily ground dwellers, and a few others as partial inhabitants of the ground. Though many opportunities opened, all three remained quadrupeds. They traveled over the forest floor and through the underbrush exactly as they had through the trees, on all fours. With hands and feet involved in locomotion, the hands were never free to use objects of any kind in defense. All found other ways of fighting off the dangers and enemies of the ground.

The patas monkey developed a decoy system and combined that with speed to outwit and outdistance a predator. If a hungry canivore approached, the dominant male patas would run away screeching and jumping about in the grass until the marauder could not fail to see him and be drawn into pursuit. With its great speed the patas could usually elude even the fast leopard. Meanwhile, the females and the young "froze" in the tall grass and went unnoticed by the predator. This adaptation to life on the ground worked well; the patas monkey survives and flourishes into the present.

The other two monkeys that came down from the trees, the baboons and the geladas, found safety in

Figure 3.1 Primates live in groups, which in turn live close to many other kinds of animals. Both social life and ecology are fundamental in primate evolution. Photographed in the Amboseli Reserve, Kenya, are: a baboon troop (above); zebras, impalas, and baboons (opposite page).

their fighting ability. The males became big enough and fierce enough with their slashing canine teeth to intimidate even some of the large cats. The male baboons also sometimes fought as a group. Sherwood Washburn watched a group of baboons and impalas feeding together, as they often did, in a grassy area near a grove of trees. He saw three cheetahs approaching; both the impalas and baboons stirred warily and uneasily, but neither fled, though the impalas were a favorite quarry of the cheetah. As the cheetahs drew close, a large male baboon stepped forward, baring his

knife-sharp canines and defying the marauders. Because the cheetahs fled, not only were the baboons protected, but the impalas also found safety. Evidently the impalas had learned from experience to rely on the baboons for protection. (See Figure 3.1.)

The Old World monkeys had minor problems of adaptation when they came to the ground. For the majority of monkeys, ground living takes place close to the trees—and this is true even for most baboons. For Old World monkeys the problem was not so much that of ground living, but rather concerned the anatomical and behavioral adaptations necessitated by living away from trees. But the Old World monkey that descended from the trees went no further in adaptations to life on the ground.

New World Monkeys

In the New World not a single species of monkey came down from the trees and survived on the ground. Here not even the primary step was taken. One reason may have been that the New World monkeys are generally small, and no very small primates have adapted to life on the ground.

Another part of the problem could be investigated in the modern laboratory. At the University of California at Berkeley, Old World and New World monkeys were placed in the same cage and closely observed as they moved over the same area and met the same obstacles. As the scientists carefully watched, they saw that the New World monkeys were less efficient than the Old in moving over the ground. Examination revealed the reason.

The lumbar region of the back of the larger New World monkey is curved differently, and it is used less when the monkey moves, and particularly when it jumps. The curvature appears to be related to the monkey's prehensile tail, which generally had been considered a locomotor adaptation, but in laboratory study turned out also to be a feeding adaptation. The monkeys use their tails, not to move faster or more surely, but to hold a branch and steady themselves while they feed with both hands. If the tail structures of American monkeys impede movement on the ground, or do not facilitate it, their tails may have precluded their successfully moving to the ground. The laboratory studies suggested that this clumsiness on the ground may have been a factor in their staying in the trees.

It had traditionally been thought that some primates were forced to come to the ground when the slow shrinkage of the forests began. But this desiccation (drying)—some 5 to 8 million years ago—occurred in the New World as well as in the Old. Instead of forcing the monkeys to adapt to a new way of life, it led to the extinction of many species. If the New World monkeys were forced down from the trees, they perished instead of learning to live on the ground.

Much more information is needed before this interpretation can be considered more than a suggestion. The situation in the New World implies that the effects of the reduction of forests on the behavior of the primates will depend on the precise nature of their arboreal adaptation. If primates are small and adapted to certain kinds of arboreal feeding, the probable result is extinction—not ground adaptation, as the desiccation theory suggests.

The prospects were narrowing. None of the New World monkeys succeeded on the ground, and the three Old World species that abandoned the arboreal for the terrestrial life stopped with this transition. To this day they have continued as quadrupeds. But the story does not end in the trees or with quadrupedal primates living on the land or partly on the land—the apes also were venturing down.

In Asia the orangutan and the gibbon occasionally climbed down from the treetops. But these apes did not often come down or remain for long, and they evolved no structural adaptations for locomotion on the ground. Their basic structure continues to be for living in trees. The gibbon swings and hurtles from branch to branch with a grace that the English naturalist William Charles Martin described as "aerial." So light and free are its movements, it scarcely seems to touch the branches in its progress. It rarely leaves the tops of trees. (See Figure 3.2.) The orangutan, on the other hand, is a slow and cautious mover; nevertheless, it is little more drawn to the ground.

One Ape Comes to the Ground

In Africa one group of apes took a different course, one that ultimately would lead to humans. This group of apes, perhaps one of the ramapithecines,

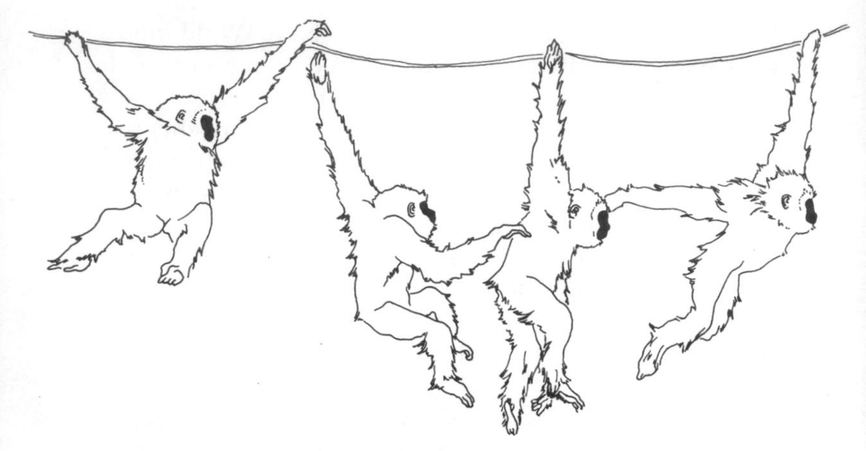

Figure 3.2 These drawings show the movements of a gibbon brachiating (swinging from hold to hold).

found life on the forest floor not only possible but rewarding. Only a few scant fossils tell of this stage in evolution. However, studies of anatomy, biochemistry, and living animals are helping to fill in this long and crucial period in history.

The apes climbing down from the tree were relatively large. Why largeness should have been an advantage is not clear, but it is a matter of record that the smaller forms were not successful on the ground.

The apes coming down for the food the land held had short, wide, shallow trunks as compared with the narrow, elongated trunks of monkeys. They also had shoulders quite unlike those of the monkeys. The ape shoulder permits a strong, powerful reach, particularly to the side. If a branch is not overly large or stable and an ape wants to reach a nearby limb, it can grab the first with the hands and swing across to a new handhold on the other. The method of swinging the dangling body to a new position, brachiation, is used by apes; most monkeys cross the same branch on all fours. (See Figure 3.3.) If a branch is sturdy enough, however, the ape also might resort to all fours or, by stand-

Figure 3.3 This drawing contrasts with Figure 3.2, as it shows the locomotion of a quadrupedal monkey (a vervet).

ing largely upright and balancing with a hold on some upper branch, might even take a few steps on two legs.

The long, mobile arms of the apes with their powerful flexor muscles also make it easy for them to climb and eat by reaching. The apes often hang from a branch by one arm and feed with the other, or they may hold a branch with one hand and stretch the other hand as far out as possible to reach a ripe fruit. A human does exactly the same thing to pick up something just out of easy reach.

There is a profound similarity in the motions of the arms of humans and apes, and on any playground one can see humans brachiating from bars, hanging by one hand, and exhibiting a variety of motions and postures that are similar to those of the apes. Humans are still brachiators, though not often in situations that call forth this behavior. Our legs are too heavy and our arms too weak for efficient brachiation, but when we climb we climb like apes, not like monkeys.

Along with this unusual structure of the trunk and arms came changes in the lower back. No backbones of these early apes have been discovered, but in some New World monkeys that also evolved brachiation the structural mechanisms of the back were similarly reduced. When these New World monkeys are studied in more detail, they should provide a much fuller understanding of the nature of brachiation and aid in the reconstruction of the locomotor patterns of the formative apes. These forms may have been in a behavioral stage very similar to that of *Ateles*—the New World monkey with the altered back.

Again the structure of living primates provides an insight into the special structure the apes of 10 million or more years ago evolved when they left the trees. The apes backing down from trees—their method of descending—could stand on their hind legs and even take a few steps upright. It was a useful ability; observers of living apes see them frequently standing to look over high grass and survey the scene for friends, enemies, or food. They are inefficient bipedalists, however, and cannot go far on two legs.

If it had not been for a modified method of locomotion they might have gone no further. But these apes coming to the ground either had or were developing another method of walking across the forest floor —they knuckle walked. Instead of placing the hand flat on the ground as monkeys did, they curled under the fingers of the hands and used the knuckles as a composite forefoot and hand. This method of walking was the key to eventual upright walking, and to human evolution. By knuckle walking, an ape on the ground could move from one isolated group of trees to another. If danger threatened, it still could climb to safety; the knuckle gait did not interfere with the structures that enabled it to climb or move around in the trees.

Although apes in zoos always had been seen standing half erect with their weight resting on the knuckles of either one or both hands, the posture was dismissed as one more peculiarity of the caged animal. The knuckle stance seemed unimportant; it was assumed that in the wilds the chimpanzee and gorilla spent most of their lives in the trees. When the anatomy of the arm was studied, it was treated as though it were an adaptation to arboreal life. The new field studies suddenly upset this neat theory. Schaller reported that gorillas are primarily ground-living animals. Nearly all their time is spent knuckle walking on the forest floor, foraging for food. Only occasionally do they climb into the trees to sun on a branch or to make a nest for an afternoon nap.

Goodall found that the chimpanzees were also primarily ground dwellers. Like the gorillas, the chimps walked through the forest on their knuckles and only occasionally walked upright. Far from being incidental, knuckle walking was the primary means of locomotion for both the great apes.

Advantages of Knuckle Walking

Knuckle walking made the transition from ape to human far easier to understand. If the apes shifting to a life on the ground had tried to walk on two feet, they would have been so awkward and slow they would easily have been run down and devoured by carnivores. With bipedalism developed in full, we are still easily outrun by lions or cheetahs. Clumsy bipeds also would have had little chance of catching the small game they hunt when they have the opportunity, or even the young that are their ordinary prey. It was always difficult to explain how a creature less fleet than a human could have survived at all. A knuckle-walking stage gets around the problem, as shown by the contemporary chimpanzee. The chimpanzee can move very rapidly quadrupedally and then climb out of danger if necessary; it knuckle walks, climbs for feeding and escape, brachiates, and may walk bipedally for moderate distances, especially when carrying something (see Figure 3.4).

If changed selection pressures (on the ground) favored the bipedal part of the behavior repertoire, then the beginnings of the human kind of bipedal walking and running might evolve while the animal could still move rapidly as a knuckle walker and escape danger by climbing. Knuckle walking got the new ground dwellers around the possibly fatal dangers of being slow, inefficient bipedalists, as most monkeys are. It also helped to solve the anatomical problem of walking on two legs instead of four. To walk on two legs in the human way, the hind legs must first combine the functions of both the forelegs and the hind legs in the quadrupedal gait.

Careful studies of the gorilla disclosed that when the huge animals knuckle walk or occasionally walk on two legs, the heel hits the ground first. This is an

Figure 3.4 Chimpanzee in Gombe Stream Reserve, Tanzania. Note that although the trunk is erect, the legs are flexed and do not extend as in human walking.

essential characteristic of human walking and is not the way monkeys walk. The monkey, and the human child learning to take its first steps, put the foot down toe first in a most uncertain and wobbly gait. Until the human child learns to use the heel, the child totters and soon sits down with a little thump. When the heel strikes the ground first, the foot does not bend, and there is power for the next step.

The gorilla foot looks like a monkey foot with the toe out at the side, but it works like a human foot. Structurally all that is needed to turn the ape foot into a human foot is to bring the big toe in line with the other toes and reduce the length of the toes. Traditional morphologists looking at the gorilla or chimpanzee foot saw it as a monkey's foot. They were not considering stability or the functional problem of the midfoot area. The more an ape's foot is studied, the more it seems like a human foot. (See Figure 3.5.)

Long arms also are essential for knuckle walking. The apes moving out on the forest floor must have had them, for their descendants do today, and many humans have arms with the same range of length.

On the ground, even part-time, apes found life very different. An ape whose hands were not totally involved in locomotion was freer to use them in other ways, a great advantage. If all food did not have to be consumed on the spot where it was found and some could be carried away, the carrier might be better fed.

A knuckle walker could carry items by curling the fingers around a small branch or a piece of fruit and holding it firmly while moving along on the knuckles. There was little interference with its gait. On the other hand, the monkey that walked on palms could not carry food.

Vernon and Frances Reynolds photographed a male chimpanzee in the Budongo Forest in Uganda knuckle walking along a road with a piece of rattan in his right hand. On several occasions they saw a chimpanzee leaving a tree carrying a fruited branch in one

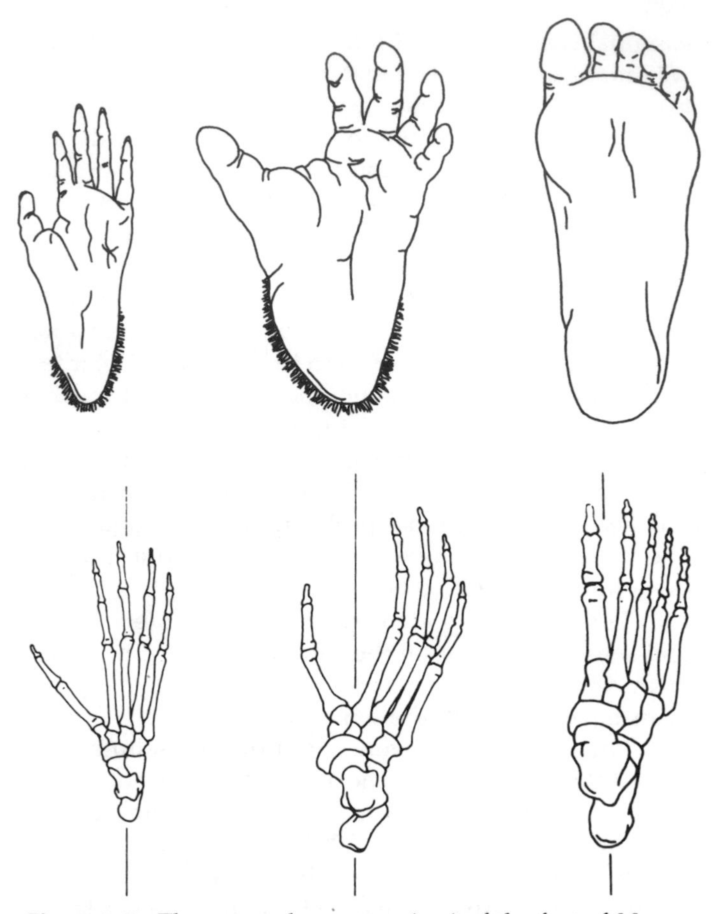

Figure 3.5 The external structure (top) of the feet of *Macaca*, chimpanzee, and *Homo sapiens* (left to right) contrasted with their skeletal structure. The lines above and below the foot skeletons show the functional axis, which comes at the same point in the chimpanzee and in humans.

hand. Once a female was observed sitting in a mahogany tree feeding on fruits still attached to a branch. She had carried it there as she knuckle walked from the fruit tree to her dining spot.

The chimpanzees of the Gombe Stream Reserve liked to walk off with blankets from the tents in the

Goodall camp, which they tore to pieces and sucked on. The blankets were either tucked under the arm or held in the hand and dragged along. One chimp was tracked as it dragged a stolen blanket for at least a mile and a half up a steep mountain. Such dragging would have been impossible for a quadrupedal animal.

The ability to walk and carry at the same time also counted heavily in one way that directly and immediately affected evolution. The infants of the chimpanzee and gorilla must cling to the mother's hair to survive, for the mother must remain active, moving around to feed, climb, and keep up with the troop. Sometimes, however, the infant gorilla needs help in clinging and may have to be assisted for as much as six weeks. Though it is difficult for her, the knuckle-walking mother can hold her baby to her with one hand and walk on her other knuckles and two legs. An observer reported that a baboon mother fell behind most of a band. Every few steps she had to stop to rest. If an adult male had not stayed beside her, stopping when she stopped and walking when she walked, she and the infant she was supporting might well have been taken by a predator. A monkey lagging behind the troop is in constant danger.

Tools and Weapons

An ape able to stand up on two legs and to use its freed hands for flailing about and throwing also is more likely to win a fight, and winning can influence both survival and reproduction. The most effective fighter has an evolutionary advantage. Goodall on different occasions observed a male chimpanzee threatening a baboon by rearing up on his hind legs and swinging his arms around above his head; the baboon fled. She also observed a chimpanzee, in a fit of rage, seize a stick and wave it in the air, then drag it behind him as it ran back and forth, when another got more bananas. Another time, Goliath, a belligerent chimp,

Figure 3.6 Chimpanzee brandishing a stick. Note how human the arms and trunk appear, but the knees are very bent and the lower part of the body is not at all in a human position.

charged a human by picking up an ax and swinging it about. (See Figure 3.6.)

When a storm was breaking or in other times of excitement, Goodall often saw the large male chimpanzees break branches off the trees and run bipedally down a hill while they waved and thrashed about with their "weapons." They would reach the bottom, return to the top, and repeat the startling, noisy performance. All the observers saw the great apes throw branches and sometimes stones. Sometimes they fired them overhand and sometimes underhand—an ability they share with humans and no other animal. The weapons were never precisely aimed and seldom hit their targets, but they frequently scared away the enemy or an unwelcome bystander.

The difference between an ape's throwing and a child's is that the ape in the wild does not aim accurately. Neither does the ape collect a pile of fist-sized stones and practice throwing them at a tree until it can hit the tree or some other target. In a zoo, however, an ape often develops remarkable skill in throwing objects at anyone it dislikes.

Because modern apes use sticks and stones, scientists believe that the early apes taking up life on the ground similarly brandished weapons.

The late K. R. L. Hall of the University of Bristol, in a review of such behaviors,[1] proposed that the use of weapons evolved in two possible ways: either (1) as a consequence of the use of larger sticks for utilitarian purposes; or (2) from the repeated discovery that a stick waved in display may do damage if it actually hits the animal against whom the display is directed and that a rock thrown in the direction of another animal during a display occasionally may hit and vanquish it. As Hall indicated, the apes did not use objects in anger alone.

Both gorilla and chimpanzee show a strong tendency to manipulate as well as throw objects, stated Jane Lancaster in her paper, "On the Evolution of Tool Using Behavior."[2] "Goodall has collected more than 1,000 of the twigs and grass blades—the tools—chimpanzees use in fishing for termites. Some will search carefully for just the right piece and then spend time preparing it. A few prepare a little pile of stems before starting to termite, and some make the twigs even before the nest is found." Along with the use of leaves for grooming, of leaf sponges for sopping up water, and of sticks for prying out roots or other foods, the great apes far surpass any other animal except humans in tool use.

To add to the difference, chimpanzees and gorillas learn their skills through watching other apes perform. A young chimp will sit for many minutes watching its mother termite, and when she quits it may pick up her fishing stalk and awkwardly try fishing. Lancaster further stated:

> The tool using behavior of chimpanzees suggests the kind of ape ancestor that might be postulated for the origin of the hominid line. It would be an ape that used tools for many different reasons and in many different ways, no matter how insignificant the tool, like leaf sponges, or how undramatic, like termiting twigs, or how inefficient, like a clumsily swung stick.
>
> The more kinds of tools the ape uses the more likely he is to be an ancestor, because it would have been the accumulated influence of many reasons for using tools that would have taken selective pressure off the specific situation, the specific tool, and the specific movement.
>
> Selective pressure was put on a hand that could use many tools skillfully and on a brain capable of learning these skills. Natural selection would then have acted on a broader category of behavior, one involving the brain, the hand, many objects, and a wide variety of social and ecological situations.[3]

Here is a perspective on the problem of learning to use objects. Primate hands evolved from claw-bearing paws early in the Eocene. For 50 million years in many families of primates, dozens of genera, and hundreds of species, hands were used for feeding, grooming, and locomotion; but in all of this time and in all of these manipulative creatures, substantial use of objects as an adaptive mechanism evolved only once.

Negative evidence suggests that a very special set of circumstances is needed to account for the beginning of tool using and that digital dexterity, although necessary (see Figure 3.7), is a very small part of the explanation. The special situation is knuckle walking.

These knuckle-walking primates had come a long way. Some 8 to 10 million years ago, one group lived in the forest. Then another group branched off. This was the chimpanzee. These animals are well adapted to the stable world in which they live. Food is plentiful and enemies are few in their forests. There was little selective pressure for change in a largely unchanging environment, so they flourished until recently. Only in the twentieth century did they encounter difficulty, when their greatly changed relative, the human, began to preempt more of their forests for human uses, such as cattle raising. Except in a few preserves, the chimpanzee now does not have a very good chance of surviving into the future.

While the gorilla and chimpanzee were beginning to go their own way, some other primates began wandering out of the forest and toward the forest edges that Simons calls the "woodland," and then on to the savannah or the more open country just beyond. Here was more space, with fewer trees, less underbrush, different foods, and different animals. Under these new circumstances, selection favored the most bipedal of the apes and those best able to fight off the speedier animals of this drier land.

As the locomotor pattern evolved in response to the new pressures, more effective tool use also evolved. Locomotion and tool use affected each other in a feedback relationship; each is at once cause and effect of the other. The use of tools was not made possible by a preceding bipedal adaptation, nor was tool use a simple discovery. Probably it was the selective result of repetitious events in thousands of populations of apes over millions of years.

As large populations of apes became more surely bipedal and as the hands were increasingly freed from locomotion, attacks with weapons must have slowly re-

Figure 3.7 To test an ape's ability to make stone tools, R. V. S. Wright let an orangutan watch a human knock off a flake with a hammerstone and then use the flake to cut a rope that secured the top of a box. Fruit was placed in the box so that the orangutan could see how a highly desired item might be reached. In the first picture, Abang is ready to begin flaking. The second picture shows the orangutan hammering at the cord, steadying the board with his foot and the back of his flexed left wrist. In the third picture, Abang is preparing to cut the cord inside the slot. The loose end of the cord hangs down from the clamp on the top of the box. With no training, Abang quickly learned to make flakes and use them, clearly demonstrating that even the hand of an orangutan is capable of making tools if the animal is given the idea. The basic problem of tool making is intellectual rather than motor.

placed attacks with the face and the teeth. The weapons proved a superior way of fighting.

About a century ago Charles Darwin argued that the large teeth of the adult male ape ancestor were replaced by tools. "The early male forefathers of man were probably furnished with great canine teeth," Darwin wrote in *The Descent of Man*, "but as they gradually acquired the habit of using stones, clubs, and other weapons for fighting with their enemies or rivals they would use their jaws and teeth less and less. In this case the jaws, together with the teeth, would become reduced in size."[4]

Darwin again was right—the fossil teeth that have been found show this evolution. The smaller teeth and the smaller jaw and neck structures that went with them (see Figure 3.8) are evidence that the apes beginning to live on the forest edges used weapons they could manipulate with their hands. If the canines had been reduced before their owners learned to fight with sticks and stones, such apes in all probability would not have survived; they would have been quite defenseless.

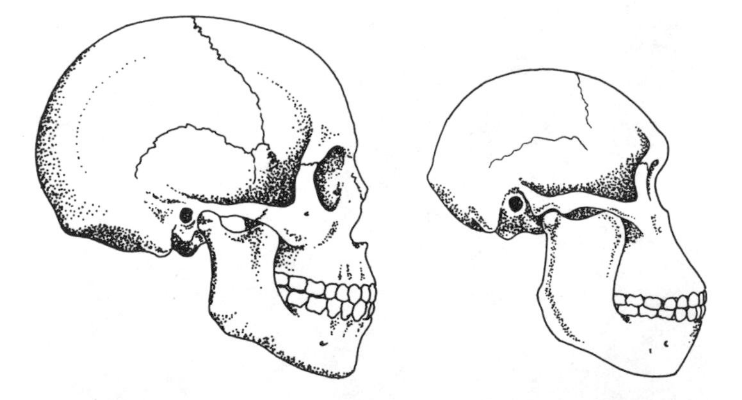

Figure 3.8 These drawings of skulls show the relatively little difference between the canines of *Homo sapiens* (left) and those of *Australopithecus africanus*.

Furthermore, a primate wandering away from the trees in pursuit of some quarry had to be ready to fight at any moment. It could not start searching for a stick or stone when it caught up with its game; the weapon had to be in hand, and it had to be used skillfully.

The skills had to be learned, and they could be learned sufficiently only in the earliest years, starting with play. To insure enough time for childhood play, the preaustralopithecines and australopithecines had to mature more slowly than their ape ancestors. Living out in the dry savannah away from the trees would have come long after ground living, knuckle walking, toolmaking, tool using, and slower maturation.

The creatures venturing out from the edges of the forest departed in other ways from their ancestral apes. They were small in comparison with humans—probably between four and five feet tall. Nor were they heavy —the 500-pound gorilla had remained behind. They were relatively slow and lacked the fighting canines of their ancestors.

To survive in the more open, drier lands they had to replace escape, size, speed, deception, and built-in weapons with other strengths and defenses. This they did successfully, as history proves. The move to the forest edges, dangerous as it was, was another beginning, not an end. They were ready with their two-legged gait, their tool use, their prolonged time for learning. They also had the anatomical and social adaptation to go further.

The bipedal way of locomotion evolved in spite of the fact that it is inefficient and costly. Humans pride themselves that a very few members of the species can run a four-minute mile on a well-prepared track—but that is only fifteen miles an hour. Wild dogs, hyenas, many kinds of antelope run well over thirty miles an hour in rough country. To hunt like a carnivore, a human female would have to run a two-minute mile carrying a baby! Primitive quadrupedalism is far more efficient than bipedalism. Further, a quadrupedal monkey or knuckle-walking ape can move quite well on three limbs.

Under past conditions a human biped could have been fatally handicapped by an injury to one leg. Even a badly sprained ankle could have been fatal unless other humans provided help and food. The more maturity was delayed, the greater the likelihood of death before reproduction and thus of elimination from contributing to the evolution of the species. Because selection was clearly in favor of a particularly vulnerable species that could move at less than one half the rate of game or predators, it is our belief that it was the success of the use of objects for a wide variety of purposes that led to the evolution of our way of locomotion.

The study of locomotion shows why it is essential to consider behavior when studying the problems of human evolution. From the study of bones alone it might be concluded that human bipeds could have run faster than apes or monkeys. It would not have been apparent that the evolutionary problem is to account for the origin of a pattern of locomotion characterized by a great reduction in speed and an increase in vulnerability.

"If the gorilla and a few allied forms had become extinct," Darwin said, "it might have been argued with great force and apparent truth that an animal could not have been gradually converted from a quadruped to a biped, as all individuals in an intermediate condition would have been miserably ill-fitted for progression."[5] Darwin did not have the field studies, the fossil discoveries, and the biochemical data to inform him, although he argued correctly from what he did know. It remained for the newer studies to show how and why such a nearly impossible transition was made by one group and one group alone.

In evolution, success is measured by reproductive success, by leaving descendants, so it is important to look at the numbers of our ancestors. Enough fossils have now been found so that estimates can be made. F. Clark Howell estimates that there were many baboons to one human. Putting it differently, humans were about as common as the carnivores in deposits of fossils.[6] Although our human ancestors surely ate more than meat, meat may have been a limiting factor. In the forests of Southeast Asia there were from three to six orangutans per square mile, but some ten square miles were required to support one human hunter. It was not until after the origin of agriculture that the numbers of humans began to equal those of the monkeys and apes. This success has been so dramatic that it is hard to think of humans as being rare compared with most primates. It is only a recent phenomenon that the tool-using biped has become numerically overwhelming.

4

MISSING LINKS

Late in 1924, as Dr. Raymond A. Dart was dressing for a wedding that was to be held in his Johannesburg home, a delivery arrived with two large wooden boxes. Dr. Dart, a professor of anatomy at the University of Witwatersrand, knew that the crates contained fossils from a lime quarry at Taung in the Bechuanaland Protectorate. The time could not have been worse, but he pulled off his wing collar and, over the protests of his wife, rushed out to see what the boxes might hold. He pried off the lid of the first box and could make out traces of fossilized eggs and turtle shells. This was not too promising. The doctor was hoping for a fossil baboon skull, similar to one previously found at the limeworks. Dart went on to the second box, and as he pulled off the lid he saw a cast or mold of the interior of a skull.

"I knew at a glance that this was no ordinary anthropoidal [great ape] brain. Here was the replica of a brain three times as large as that of a baboon, and considerably bigger than that of any adult chimpanzee. . . . But it was not big enough for primitive man."[1] Was there a face to go with the brain? Dart ransacked through the rocks and found a large stone with a depression into which the cast fitted perfectly. Faintly visible in it was the outline of part of a skull and a lower jaw. The face might well be somewhere in the block. At this point the bridegroom arrived; he was Dart's close friend and Dart was to serve as the best man in the wedding. So Dart had to tear himself away from the boxes.

The moment the last guest departed, Dart was back at his boxes and the puzzling skull. In fact, every possible minute during the next seventy-three days was spent trying to free the skull from its matrix. On the seventy-third day the stone parted. What emerged was a child's face, with a full set of milk teeth with the first permanent molars just beginning to erupt. (See Figure 4.1.)

Was this child ape or human? If it were an ape, Dart argued to himself, what was an anthropoid with a brain larger than that of a chimpanzee and as large as a gorilla's doing in South Africa, more than 2,000 miles from the jungles in which the great apes lived? Dart knew that there had been little climatic change in South Africa in the last few million years. The dry, treeless veldt of the Transvaal would have offered none of the foods on which the great apes normally feed.

The skull was shaped like that of a human, and although it was low, it had a true forehead and none of the ape's heavy eyebrow ridges. The canines were small, too—not like those of an ape. As he studied the position the skull must have occupied on the body, his

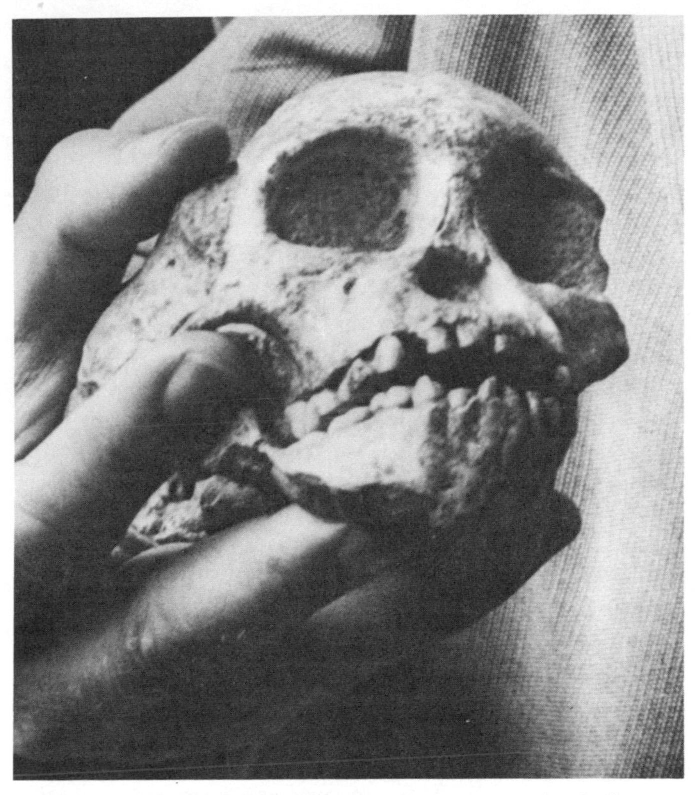

Figure 4.1 The hand holding the Taung juvenile skull gives an indication of its size.

conviction grew: this creature emerging from the block of stone had walked upright.

It was, by all the signs, a humanlike ape rather than the apelike man of fiction. But Dart was cautious. He named it *Australopithecus africanus.* Dart concluded that the skull was intermediate between living apes and human beings.

Once again there was a furor. Sir Arthur Keith, the leading authority on fossil humans and conservator of the Huntarian Museum in the Royal College of Surgeons, conceded that the Taung find was a "remarkable" ape but held that it should be placed in the same group or subfamily as the chimpanzees and gorillas. Keith added that it seemed to be "near akin to both." Other anthropologists expressed doubt or complained about the "barbarous" name Dart had chosen.

Newspapers and magazines widely reported the "finding of the missing link," and the *Spectator* ran a contest for the best epitaph in verse or prose. The classic music hall joke was given a new twist—"Who was the girl I saw you with last night—was she from Taung?" The Sunday *London Times* printed a letter from a "plain but sane woman" asking Professor Dart, "How can you, with such a wonderful gift of God-given genius become a traitor to your Creator by making yourself the active agent of Satan and his ready tool?"

The press also carried a story from the United States: "Professor Dart's theory that the Taung skull is a missing link has evidently not convinced the legislature of Tennessee. The governor of the state has signed an anti-evolution bill which forbids the teaching of any theory contrary to the Biblical story of creation or that man is descended from any lower orders of animals."

Among the relatively few messages of congratulations was one from Robert Broom, a Scottish-born physician who had been drawn to South Africa in his search for fossils. Broom was seeking the answer to another debated question of the day: Did the mammals arise from an amphibious or a reptilian ancestor? Broom did an intensive study and promptly wrote an article for *Nature* in which he said: "In *Australopithecus* we have a being with a chimpanzee-like jaw, but a sub-human brain. We seem justified in concluding that in this new form discovered by Professor Dart we have a connecting link between the higher apes and one of the lowest human types."[2]

After the initial flurry the world's interest in the Taung skull faded rapidly. The anthropology headlines were being made by the search for man's earliest ancestors in Outer Mongolia and in the caves at Choukoutien near Peking. It was confidently anticipated that Asia would prove to be the cradle of humankind. Dart summed up the attitude: "What dissonant squawkings were these from the puny South African infant at Taung? What was the use of a baby anthropoid when you were looking for men—primitive men at that?"[3]

Dart and Broom realized that only the discovery of an adult *Australopithecus* would overcome the world's condescending doubts and joking unacceptance. Broom said that Dart felt hurt and discredited by the reception of his discovery. Dart himself emphasized that as dean of the Faculty of Medicine he had neither the time nor the inclination to continue the search for the necessary adult form.

Neither Dart nor Broom gave up their intense interest in the skull itself. During the next four years they continued their efforts to free the skull completely from its matrix and finally, in 1929, succeeded in separating the upper and lower jaws. This made it possible for the first time to see the full pattern of the teeth. When Dart sent casts to the world's authorities, William K. Gregory, curator of comparative anatomy of the American Museum of Natural History, was convinced: "In the light of all this additional evidence if *Australopithecus* is not literally a missing link between our older dryopithecoid group and primitive man, what conceivable combination of ape and human characters would ever be admitted as such?"[4] The details of the teeth often provide strong evidence for proving relationships. The teeth of the Taung skull are humanlike and very different from those of the apes.

Broom, unlike Dart, was eager to take up the search for more material. However, he was tied to his medical practice in the town of Maquassi, inconveniently far from both Taung or any other likely fossil sites. Jan Smuts, the scholarly prime minister of the Union of South Africa, decided to do something about this unfortunate situation, and he appointed Broom curator of vertebrate paleontology in the Transvaal

Museum in Pretoria. "Gen. Smuts thought it a pity that I should be spending my latter years in medical work when I might be devoting all my time to scientific work," Broom said.[5]

Broom was then sixty-eight, and in 1936 when he was ready to start the field search, he was nearing seventy. He soon unearthed some unknown species of rats and moles and an ancient baboon skull. When some students of Dart's told Broom about some baboon skulls they had found in another limestone quarry at Sterkfontein, Broom quickly arranged a trip to the site. The town, about thirty miles from Johannesburg, had grown up during the gold rush of 1886, and in the intervening years millions in gold had been taken out, as well as a wide array of fossils. Youths were always picking up the fossilized bones of extinct baboons and other animals. The town had even published a guidebook urging visitors to "Come to Sterkfontein and find the missing link."

Broom and the students arrived on August 9, 1936, and went first to the limeworks quarry. The manager, G. W. Barlow, knew about the Taung skull and agreed at once to keep an eye out for any other fossilized bones his quarrying might blast out. He "rather thought" he had seen some skulls of the kind the doctor described. Usually he sold any "nice bones and skulls" to Sunday visitors to the quarry.

When Broom returned the next week, the manager handed him about two-thirds of a fossel brain cast, which Broom recognized instantly as that of an adult australopithecine skull. It had been blasted out only that morning. Broom rushed over to the pile of debris that had been left and began searching for the skull that had made the impress. After hours of digging and sorting through the dusty pile, Broom found the base of the skull, the upper jaw, and fragments of the cranium. When all the bits and pieces were cleaned and assembled, he had a nearly complete skull of an adult with nearly all of the half-human, half-ape characteristics that the world had considered a freak or an impossibility in the Taung skull.

Broom continued to visit his fossil "gold mine" at regular intervals. All the workers were alerted to watch for any fossils and always had a stack awaiting his inspection—bits of skull, teeth, and quantities of animal fossils. "Every trip cost me some shillings in tips, but it was worth it," Broom remarked.

One morning in June 1938, when Broom arrived, Barlow met him with an air of promise. "I have something nice for you this morning," he said, handing him an upper jaw of a hominid. The manager had obtained it from a schoolboy, Gert Terblanche, who lived on a farm at Kromdraai about two miles away.

Broom jumped in his car and rattled off to the Terblanche farm. Gert was at school, but Broom succeeded in getting him excused from his classes by agreeing to deliver a talk to the students on the fossils in their area. Gert led Broom to a cache where he had secreted his finds and presented him with a "fine jaw with some beautiful teeth." During the next few days, he and the boy recovered many additional scraps of bone from the place where Gert had dug out the jaw. When the pieces were all fitted together, they constituted most of the left side of a skull as well as the lower right jaw.

The new skull was surprising and puzzling. It showed a mixture of apelike and humanlike characteristics, but it differed notably from *Australopithecus*. The jaw was more massive, the grinding teeth larger, and the face flatter and more apelike. In their form, however, the teeth were distinctly human. Broom was convinced that the being he and Gert had screened from the rock of a former cave was so different from the Taung skull, or *A. africanus*, that it would have to be assigned to a genus of its own. He named it *Paranthropus robustus* (robust near man).

The claim that not one, but two genera of hominids had existed in one small area of South Africa further increased the disbelief of the English authorities. "Of course the critics did not know the whole of the facts," said Broom. "When one has jealous opponents one does not let them know everything."[6] It must be remembered that the early excavations did not recognize that the cave deposits were mixtures of remains from very different ages. It is now known that the levels with the fossils of *Australopithecus* are older than the levels with the stone tools.

At the time Broom had no accurate way of dating the fossils he was unearthing or the geological formations in which they lay. At Sterkfontein the caves in which the hominids had lived had filled gradually with dense layers of water-deposited travertine. The forming stone surrounded and covered the bones that lay in the caves. In time the caves filled in, and often the roofs collapsed. The modern quarrying operations exposed the filled-in former caves.

At Kromdraai, erosion of the surrounding earth had left the hard consolidated cave deposits as small kopjes, or mounds. The masses of bones in the former caves were compacted into a conglomerate called bone breccia. But at neither site was there any clear stratification to give a clue to time. Only general estimates of the age of the caves and their contents could be made.

The outbreak of World War II forced Broom to stop his work in the field. By that time he had accumulated much material, though in the rush of collecting it he had not had time to prepare and study it thoroughly. The war years gave him that opportunity, and in 1946 Broom and G. W. H. Schepers were able to publish a full report, *The South African Fossil Ape-Man*.

The opposition began to crumble. The South African discoveries could no longer be brushed aside. Sir Wilfrid E. Le Gros Clark, who had been a skeptic, was so impressed with the data presented in the Broom book that he arranged to visit South Africa in 1947. Before he went, he carefully examined the skulls and teeth of more than a hundred modern anthropoid apes to determine the normal variations in their anatomical structure so he could compare these measurements with those of the australopithecines. The South Africans gladly opened their collections to him.

"The results of my studies were illuminating, not only because they made me realize how much more profitable it is to study original specimens than to rely on casts or photographs, but because they convinced me that Dart and Broom were essentially right in their assessment of the significance of the Australopithecines as the probable precursors of more advanced types of the Hominidae," Sir Wilfrid said publicly, at the First Pan African Congress on Prehistory held shortly afterward at Nairobi.[7]

He had a battle on his hands. Professor Wood Jones, a distinguished anatomist, argued that humans had evolved, independently of monkeys and anthropoid apes, from the tarsiers or tarsierlike animals, and had come into existence about 60 million years ago. He therefore rejected the idea of such "missing links" as the australopithecines with their combination of apelike and hominid characteristics. Additional finds and further studies slowly converted most of the other doubters.

At the end of the war, though Broom was eighty, he was eager to resume the search for more australopithecines. Smuts offered him all the support he might need. However, when the prime minister left the country on a long trip, the Historical Monuments Commission informed Broom that he could do no more work without a permit and that a geologist would have to be present to make certain that no damage was done by any blasting. Broom was outraged. Smuts, upon his return, apparently agreed with Broom and told him to "carry on."

Broom and his assistant, John Robinson, then resumed work at Kromdraai. When a permit arrived three months later, it said nothing about operations at Sterkfontein. Broom's answer was to shift operations to the quarry. Within a few days he had some fine teeth and part of a child's skull. Two weeks later a blast loosened a large chunk of breccia that at first looked unpromising. As the smoke cleared, however, a whole, perfect skull stood revealed. "I have seen many interesting sights in my long life," said Broom. "This was the most thrilling of my experience." The blast had split the skull in two, and lime crystals lined the braincase. Even before the skull was cleaned and put together again—a minor matter—Broom pronounced it "the most valuable specimen ever discovered." It would have been difficult to prove him wrong. Broom bestowed a special name and rating on his find, *Plesianthropus*, and the adult female represented soon became known around the laboratory as Mrs. Ples.

Public excitement about the unusual find ran high. Mrs. Ples was not only a remarkably clear emanation of a long-unsuspected past; her discovery had also broken the law. The doctor's flouting of the commission had become too obvious to ignore. An investigation was called for and made. It established that there were no rock strata at the place where Broom was working and that no damage had been done. Broom was issued a permit: "So they had to allow me to continue though under still absurd conditions to which I pay no attention," Broom announced.

Broom felt he was justified when soon thereafter he found a nearly perfect male jaw and, on August 1, 1948, a complete pelvis. Usage had ground down the canine tooth in the jaw until it was in line with the other teeth, exactly as in humans. Similar wear is never found in the projecting fanglike canines of the male anthropoids.

The pelvis settled the most critical problem. It showed beyond all question that these hominids had walked upright, or nearly upright. Up to that time a few bits of leg bone indicating bipedalism were dismissed as human bone that had somehow become mixed with the ape remains. What was equally significant, for all of its human characteristics, was that the pelvis was not entirely human. "What we can now say is that the pelvis is not in the least anthropoid, and that it is nearly, but not quite human," Broom concluded.

Soon after, a University of California expedition arrrived in South Africa and offered to finance the work if Broom would open up a new cave. A site at Swartkrans, just across the valley from Sterkfontein, was selected. True to form, within a few days Broom found another massive jaw. The teeth were so large and square, though they were human in conformation, that they suggested comparison with some of the giant teeth that had long been found in China.

The world's honors as well as its recognition were coming to the intrepid and dedicated Dr. Broom. The Royal Society of South Africa held a major international scientific congress in his honor and published a Robert Broom commemorative volume. Many medals and degrees were conferred on Broom in the United States and Great Britain. His life was at its fullest when he died on April 6, 1951, at eighty-four.

Le Gros Clark

Professor Le Gros Clark had continued the studies he began in South Africa. By 1949, when he prepared a formal appraisal of the australopithecines for the *Yearbook of Physical Anthropology*, the remains of some thirty individuals had been found (see Figure 4.2). The young and the old were represented, males and females, and enough parts of the body for a reconstruction of the whole. Never before had anthropology had such a wealth of material.

In his summary Le Gros Clark said:

> From all of this material it is evident in some respects that [the australopithecines] were definitely apelike creatures with small brains and large jaws. Indeed, in the general proportions of the brain case and facial skeleton they represent a simian level of development

Figure 4.2 The African sites of major finds of *Australopithecus*.

> not very different from that of the modern large apes. [See Figure 4.3.]
>
> But in the details of the construction of the skull, in their dental morphology, and in their limb bones, the simian features are combined with a number of characters in which they differ from recent or fossil apes and at the same time approximate quite markedly to the Hominidae.

Clark further held it "a reasonable inference" that the "astonishingly primitive hominids" either were in the main line of human evolution or were only slightly modified descendants of such a group.

Now virtually all doubts were resolved. One who yielded completely was Keith:

> I was one of those who took the point of view when the adult form was discovered it would prove nearer akin

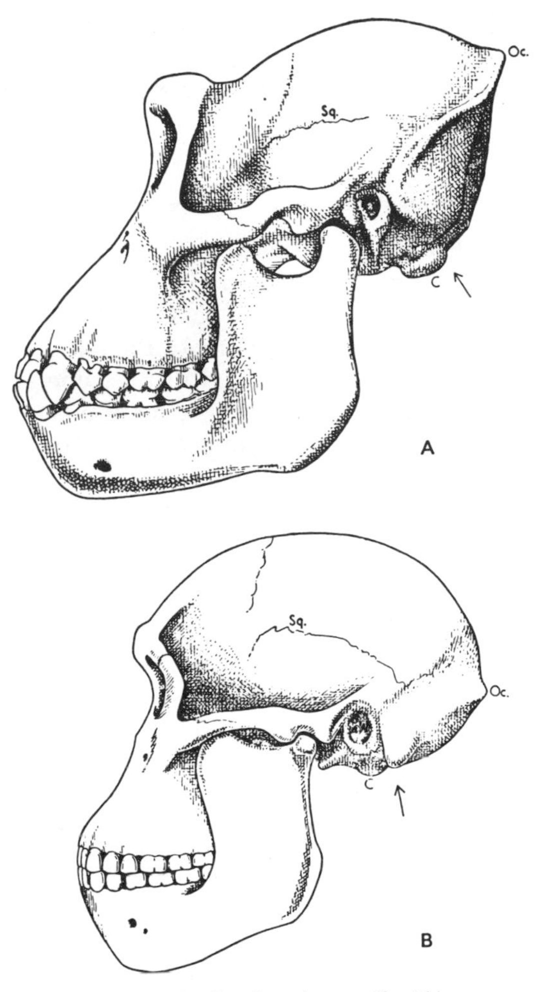

Figure 4.3 The skull of a female gorilla (A) compared with that of a fossil, *Australopithecus* (B). Note that in *Australopithecus* the anterior teeth are much smaller, as are all the parts of the skull associated with the jaw muscles.

to the living African anthropoids, the gorilla and the chimpanzee. . . . I am now convinced on the evidence submitted by Dr. Robert Broom that Professor Dart was right and I was wrong. The *Australopithecinae* are in or near the line which culminated in the human form.[8]

By all the evidence these appeared to be creatures with apelike brains and humanlike bodies. The exact reverse had been anticipated; for some it remained a shock and unacceptable.

Tools—Absence and Presence

A negative finding raised one problem for the South African investigators. In the twenty years of their intensive search no stone tool of any kind had been found in association with any of the fossils. Stone tools from later periods abounded in the general area, but it began to look as though these hominids had not yet acquired the skill to make stone tools. Many argued that this was to be expected because it was assumed that they lacked the mental capacity to conceive of tools or to work stone.

With the aid of money from several foundations, Dart made a detailed analysis of 7,159 fragments of bone, tooth, and horn taken from the Makapans breccia in one year of work. About 92 percent of the bone came from various types of antelope. The australopithecines' favorite meat seemed to have been venison. As Dart continued with his sorting and compiling he found sharp-edged pieces of bone that he thought had been used for cutting and scraping. On the basis of his analyses of the fragments Dart concluded that a Bone Age had preceded the Stone Age.

Once again he had created a controversy. Many of his fellow scientists questioned whether the australopithecine "bone tools" were actually tools or only bones broken by the gnawing of hyenas and other animals that had hauled the carcasses into their caves. One investigator collected an assemblage of bones from a modern hyena cave. An analysis showed that the bones had been broken and shattered in almost the same way as Dart's bone tools.

In 1954 C. K. Brain asked Dart's permission to make some soil tests in his dig at Makapans. When

Brain returned from the cave he showed Dart 129 chipped, used, or damaged stones he had found in an eighteen-foot sandy layer only twenty-five feet above the australopithecine-carrying gray breccia. The two immediately took them to geologist van Riet Lowe. Lowe spread them out and soon separated 17 pebbles from the others. After careful study he told his co-workers, "I'm absolutely satisfied that these are pebble stone tools of the type I've already described from the highest gravel terraces of the Kafu and Kagera Rivers" (Uganda).[9]

Only a few chips had been removed from the fist-sized stones. The rest of the stone was untouched, and only close scrutiny disclosed that the broken edges were not a normal breakage. Chips had been removed and a usable edge created by a being who had acted deliberately and purposefully.

In a radio address that followed the announcement of the finding of the tools so close to the australopithecine layer, Professor Lowe said: "The discovery of the oldest recognizable man-made stone implements in deposits immediately over remains of manlike apes is of the greatest significance. It narrows the gap between ape and man as it has not been narrowed before."[10]

While numerous hominid fossils were being found in South Africa, Louis and Mary Leakey were carrying on the search for those forerunners in East Africa; it was a long search and for many years a futile one. Leakey had been born in Africa; the son of English missionaries to the Kikuyu tribe, he learned Kikuyu as soon as he did English. After attending college in England where he specialized in prehistory, he returned to Africa in 1924 and soon became director of the Coryndon Museum in Nairobi.

He was only 500 miles away, but it took a week-long drive over roadless wilds to reach the Olduvai Gorge. This 25-mile long gash in the earth—a part of the Great Rift Valley—was without much question one of the world's great repositories of fossils. (See Figure

4.4.) Wilhelm Kattwinkel, the leader of a German expedition to Africa who had discovered it by nearly falling into it, had reported that fossils were present in an unimaginable profusion. He sensed that the 300-foot-deep walls were a book of life. The record was there, from the beginnings at the floor of the gorge to the life of today on the Serengeti plains stretching back from the rim. When World War I halted German explorations and postwar difficulties prevented their resumption, some German scientists urged Leakey to take over. Leakey needed no urging—he was more than eager to go, but until 1931 he lacked the financial backing for an expedition.

At Olduvai, during the dry season, water had to be carried from a spring some thirty-five miles away on

Figure 4.4 A view of Olduvai Gorge.

the edge of the Ngorogoro Crater. At night Leakey counted the eyes of eleven lions shining in the darkness around his tents: "We never bothered them and they never bothered us,"[11] he always explained. But fossils were everywhere—a giant pig as big as a modern rhinoceros, pygmy animals, all animals long extinct. Ultimately Leakey counted more than a hundred extinct species. But the way to search for early humans was largely by crawling on hands and knees along the rough sides of the gorge to look for scraps of bone or tooth.

It was possible to work in the semidesert area only in the dry months—in the wet season downpours made the gorge a quagmire—and Louis and his wife, Mary, could seldom afford to remain for more than seven weeks a year. But the Leakeys, accompanied in later years by their sons, Richard and Jonathan, went back year after year.

Meanwhile, in South Africa one find after another was being announced. The scientific fanfare was great, but for nearly the first thirty years the South Africans had found no stone tools with their hominids. It was frustrating. The Leakeys were finding numberless stone tools ranging from the simplest to the most skillfully made, but they could find no sign of the makers of those simple early tools. It began to look as though early humans had left their tools in one part of Africa and their bones in another.

On one memorable day in July 1959, Leakey developed a fever and had to remain in camp. But the day could not be lost, so Mary, accompanied by two of the Dalmatians that always stood guard against snakes and intruding rhinos, went out to the place where they were working. On the hillside a bit of bone and glint of teeth caught her eye. Two teeth, brownish gray and almost iridescent, were just eroding from the rock. The teeth were nearly twice as wide as ours, but they were clearly human in shape. After photographs were made, the Leakeys went to work with camel's-hair brushes and dental picks and found a palate lying behind the teeth and a partial skull. The expansion and contraction of the rock had cracked the fossil into more than 400 fragments. Some pieces were recovered by screening all of the scree, or fine rock debris, that lay around and on the slopes just below. The work took nineteen days.

While the task of putting the skull together went ahead—Louis compared it to a complex three-dimensional jigsaw puzzle—the Leakeys continued to excavate the area. A wholly unexpected scene was revealed. The skull had lain on what proved to be an ancient living floor or campsite on the edge of a lake that had long since disappeared. Lying about on the former beach were the bones of numerous small animals the campers had undoubtedly eaten, and nine of the pebble tools with which they probably had killed or skinned their prey. Even some of the chips that had been knocked off the rocks to make their jagged, sharp edges lay scattered here and there. Later many more tools were recovered.

The Leakeys had seen hundreds of such simple tools on the floor of the gorge and in its lowest strata, but never before—either at Olduvai or in South Africa—had these beginning tools been so closely associated with their probable makers. Doubts that early humans could have been toolmakers were immediately lessened by this ancient scene.

Leakey permitted his scientific imagination to reconstruct what might have happened on the day that somehow had been frozen into stone for millennia to come. Perhaps rain had fallen for many days and the lake level was rising ominously. The band must have known that by dawn they would have to retreat. As they went to arouse an eighteen-year-old youth who had been ill for some time, they found him dead. Perhaps the hard-pressed campers covered his body with brush to protect it from the hyenas and then fled to higher ground.

The rising water may quickly have covered the boy's body, the tools the hunters left behind, and the bones from their recent meals. A thick layer of silt from the muddy water settled over the whole scene. In the years that followed, other rises of the lake deposited more and deeper silt on the mud-encased floor. Finally the lake itself vanished, leaving the boy and the campsite entombed under several hundred feet of sediments that were hardening into rock. Occasionally, volcanoes threw their ash and lava across the area.

The campsite might have remained forever buried if earthquakes had not convulsed the area and split it with a great crack that eroded into the gorge. The torrential rains that came at times even in that dry land continued to erode the sides of the canyon until at last the ancient campsite and its dead lay partly exposed at the face of the rock. If the Leakeys had not come along when they did, the fossils and all the evidence of that lost day would soon have been washed anonymously into the gorge floor below.

The Leakeys tallied the animals on which the campers had fed—frogs, birds, a tortoise, some young pigs, a juvenile giant ostrich, rats, mice, lizards, and snakes. The selection indicated that even with their sharp-edged pebbles the campers probably had been unable to attack giant- and medium-sized animals of that time.

As the skull took shape—only the lower jaw was missing—the Leakeys saw an australopithecine type of being. Here was proof that early humans had spread over large sections of Africa, for the gorge lay more than halfway up the eastern part of the continent from the australopithecine territory in the south.

But the skull and the face looking out at the reconstructors was much more like the big, massive hominids found originally at Swartkrans than the small forms first found at Sterkfontein or Taung. It too was certainly robust. The huge molars were at least twice the size of modern grinding teeth; they could have cracked nuts, and the Leakeys sometimes referred to their discovery as "nutcracker man." Detailed examination confirmed the humanlike shape of the teeth. The canines and incisors were reduced as they were in *Paranthropus robustus*, and the third molars were relatively small as in *Australopithecus*.

The skull was low, as in the other hominids. But Leakey emphasized that the curve of the cheek and the general facial architecture showed advance toward *Homo*. He set up a separate genus and named the fossil *Zinjanthropus boisei* (*Zinj* means East African in Arabic, and the *boisei* was for Charles Boise of London who helped finance the expedition). Leakey estimated that *Zinj* might be about 600,000 years old. His guess was based only on the generally accepted age of the period to which some of the animals belonged. It was about this time, however, that other precision methods of dating the past were being developed. Radioactive carbon, or carbon 14, had been developed first and had made possible the exact dating of organic remains that were not more than some 50,000 years old. A longer-term clock was developed by using radioactive potassium and measuring its decay into argon.

"If you were to carry 18 potassium atoms around in your pocket for approximately 1.2 billion years, you would find upon counting them that only half were left," said Garniss Curtis of the University of California, who with Jack Evernden was an early developer of the method. "In place of the missing potassium atoms you would find eight calcium atoms and one argon atom."[12]

The lava that had flowed over the *Zinjanthropus* site had contained potassium, as nearly all lavas do. Would it also contain argon? The California scientists went to Africa to collect samples and made measurements showing that the lava flow had occurred about 1,750,000 years ago. *Zinjanthropus* and the beach camp were not 600,000 years old, but 1,750,000!

Meanwhile the Leakeys, with increased aid from the National Geographic Society and foundations, were expanding their operations. In a stratum about a foot below the *Zinj* level they made another discovery of first importance. They found a fossilized left foot that had walked on the earth at least 1,750,000 years ago. It was the first direct evidence in the foot of how humans had evolved from quadrupeds to bipeds. The foot was nearly complete, with five toe bones—only the toe tips were missing—the five middle foot bones, an ankle bone, and part of the heel.

The foot was human. These creatures of the past undoubtedly had brains no larger than those of the ape; but their feet, and the posture these indicated, were largely those of humans. This discovery confirmed the deductions made from the pelvis found in South Africa. "It helps to confirm Charles Darwin's theory of the evolution of the human race," commented F. Clark Howell, in reference to Darwin's theory that the upright stance and the development of the feet came first.

The ancient beach deposit where the foot bone had somehow survived the ages also held six finger bones, two ribs, and fragments of a skull of a ten-year-old. Further testifying to the great age and rarity of the site was a wealth of animal fossils. Some, Leakey reported, came from genera and species unknown until now. Others were from species he had not previously seen at Olduvai.

In the early 1960s the Leakeys found parts of several skulls, teeth, bone fragments, tools, and another living floor in the same area. The child from the "foot" level and the others, Leakey decided in collaboration with P. V. Tobias and J. R. Napier, had a larger brain than the massively built "nutcracker" *Zinjanthropus*, or *A. boisei*, and named the youngster *Homo habilis*. Leakey had initially assumed that the tools on the campsite were made by the *Zinjanthropus*, or *A. boisei*, band. After finding *Homo habilis*, he proposed instead that they might have been made by hominids with larger brains. That would have put two species, one more advanced than the other, in the same area at the same time.

More remains of the creatures of the distant past continued to come from the Leakeys in East Africa (see Figure 4.5) and from all the South African sites, Sterkfontein, Kromdraai, Swartkrans, and Makapans. A sizable population that had lived over several million years was emerging from the ancient caves and campsites.

Alan Mann of the University of Pennsylvania studied fossil heaps collecting in the South African area. To determine exactly how many individuals the bones represented was difficult, for some of the fragments might have come from one or from several individuals. On the basis of the most careful appraisal he arrived at these figures:

	Maximum	*Most Probable*
Sterkfontein	40	25
Swartkrans	90	75
Makapans	12	7

The totals, however, change rapidly and constantly. None of the sites is more than partially investigated, and the work continues.

What is important is that many individuals are represented and that they come, not from one, but from five sites in South Africa. The number of specimens might be compared with the eight from Java and the eleven from Peking.

Classification

As the material increased, the scholars began to classify and compare. Many skulls that seemed unique to their discoverers were held by the classifiers to be only variations in a general population. The verdict of the scholars was that all the hominids belong to two species. Le Gros Clark described the two species: *Aus-*

Figure 4.5 A skull of a fossil human *(Australopithecus)* found by Richard Leakey in the area east of Lake Turkana, Kenya. The scale shown is in centimeters.

tralopithecus africanus, a small general type; and *Australopithecus robustus,* the heavily built one. "As an approximate parallel," he said, "one may refer to the pygmy chimpanzee and the more robust common type of chimpanzee that live on either bank of the Congo, or to the forest gorilla and the mountain gorilla that are nearly neighbors in Central Africa."[13] Clark also compared the two with modern humans and with the heavier, more massive-jawed Neanderthal who flourished and then became extinct, both of which are *Homo sapiens:*

> The skull of the robust type shows quite pronounced contrasts with the gracile [slighter] type. For example, the brow ridges are more heavily built . . . the jaws are more massive, and the grinding teeth are large in size. The skull is also characterized, like that of the gracile type, by many of the same hominid features. The parallel with Neanderthal man and the modern type of man

> is certainly very apt. . . . It is a matter of discussion whether these two australopithecine types are to be regarded taxonomically as separate distinct species, or whether, after all, they are no more than two distinct subspecies or varieties of the same species.

The australopithecines, all the evidence shows, had brains in the 450 to 550 cc range (the modern human brain is 1,100 to 1,500 cc; the gorilla's, 350 to 650 cc). They walked upright. They flourished over wide, dry areas, very different from the jungles of their ancestors. They certainly had ranged from South Africa to East Africa. Did they go farther?

Ethiopia

There was no answer until the late 1960s. About a thousand miles north of Leakey's dig at Olduvai is a fossil-rich area in Ethiopia. In a low area at the north end of Lake Turkana, fossil bone lay scattered thick around the dry dusty surface and protruded from the faces of the low hills. An early report was made in 1891, and in 1896 an expedition verified the fossil wealth.

Far in the past the Lake Turkana system had been connected with the Nile system—some of the fish remain the same into the present time. After separation, the lake and the Omo River that fed into it often overflowed. In other eras the waters had retreated, but a rich animal life existed on the beaches and in the swamps. With its game and water, the area might also have been one to attract early hominids. Scientists were eager to investigate one of the great nearly unexplored fossil banks of the world.

Until the 1960s Emperor Haile Selassie refused all who wanted to go in, but he followed the discoveries in East and South Africa. Leakey finally was able to convince him that Ethiopia too should be opened to science; and Haile Selassie agreed to welcome an expedition, the International Palaeontological Research Expedition, to the Omo Valley. It was made up of French, Kenyan, and U.S. teams under the respective leadership of C. Arambourg, Richard E. F. Leakey (son of Louis and Mary Leakey) and F. Clark Howell of the University of California. Work began in 1967.

A camp was set up near Lake Turkana. By 11:00 A.M. the temperature was 108 to 110 degrees. The only way to describe the heat was "hotter than hell." The scientists worked from dawn to early afternoon and from late afternoon to dark. A shuttle plane brought supplies from Nairobi—about 750 miles away—and a helicopter assisted in the mapping of the area and in locating promising places. Areas of potential interest seemed to be almost everywhere. The expedition listed more than a hundred localities with fossils, all except two in an area of about twenty square miles. "Many of the fossils were exposed at the surface," said Howell. "If we did not get to them they would disappear in the next big rain."

In two years of work more than eighty species of animals were identified: elephant, rhinoceros, hippopotamus, giraffe, lion, monkey, rodent—and humans. Perfectly preserved teeth, some quite unworn, were found at the White Sands and Brown Sands locations, as well as many pieces of fossil bone.

The teeth and bones the scientists were scratching from the dry Ethiopian earth were, to their great interest and reward, australopithecine. So *Australopithecus* had been in Northeast Africa (see Figure 4.6) as well as in the east and the south. The range—and the story of evolution—was being vastly expanded.

The hot Ethiopian land was yielding even greater surprises. Some of the teeth seemed to be close in size and detail to Leakey's *Homo habilis*. Others resembled *A. robustus;* still others, *A. africanus,* the small gracile form originally found at Taung and Sterkfontein. One tooth, which had just erupted when its owner died—only a tiny patch of wear was shown—had erupted in a pattern that essentially duplicated the tooth of *A. afri-*

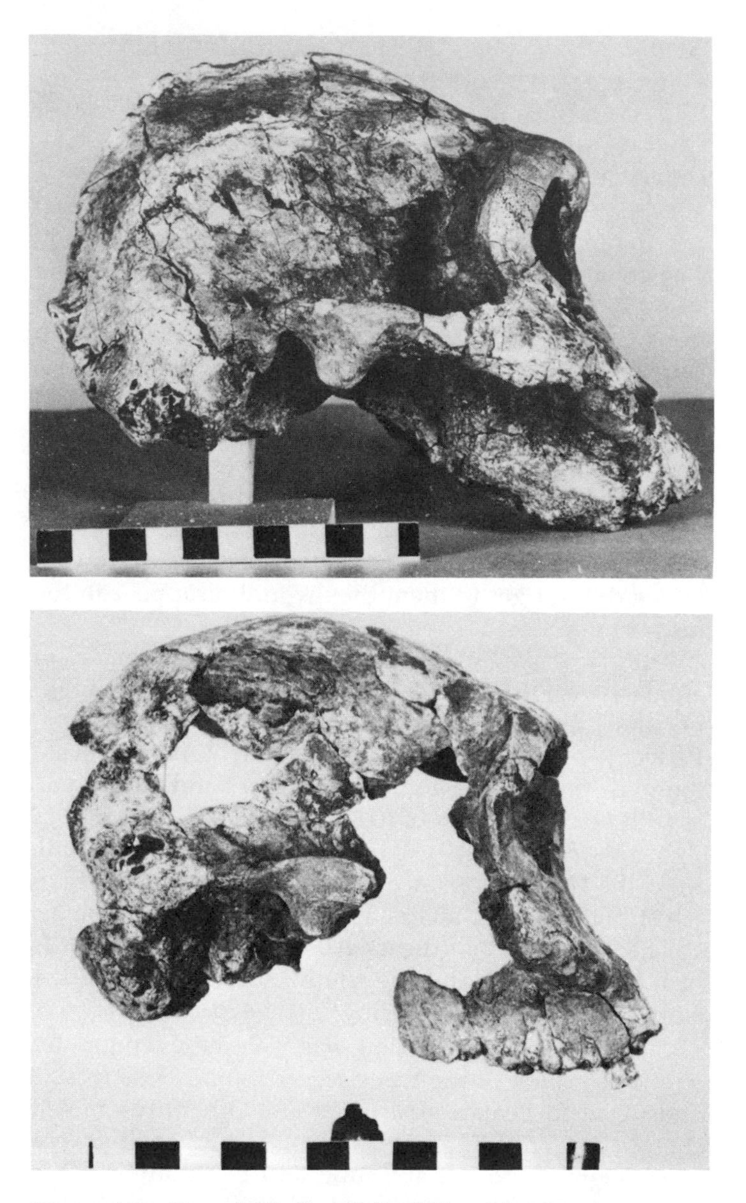

Figure 4.6 Two skulls found by Richard Leakey in an area east of Lake Turkana, Kenya. These skulls probably represent the male (top) and female of *Australopithecus*.

canus. In every case, however, there were differences as well as resemblances to the hominids in other parts of Africa. Until the expedition could find more material, Howell suggested that no final attempt be made to determine the species of the hominid teeth and bones.

The river and lake deposits that held the australopithecine remains had been overspread several times by volcanic ash and lava, which meant that potassium-argon dates could be obtained. Samples sent to the dating laboratories yielded exciting reports—one stratum had been laid down nearly 3 million years ago. Just below it lay some teeth and a jawbone fragment of the early inhabitants. The hominids of the Omo, then, were at least 3 million years old; and some nearby beds showed dates of about 4 million years.[14]

Suddenly the time set for the first creatures to leave the forests and make their way into the open lands was nearly doubled. The evolution of humans was placed in a new time perspective, and the scientists began new appraisals and calculations.

If the australopithecines were spread from South Africa to Northeast Africa at before 3 million years ago and if the dates of their separation from ape to humans were set at 5 to 10 million years, the gap from ape to humans was growing very narrow. Few links were any longer missing. What humans had been at nearly every stage could now be seen. From ape to human—it was a steady and little-interrupted progression, accomplished in a few million years.

Both climate of opinion and the order of discovery have influenced the interpretation of human evolution. Suppose that the footprints found at Laetoli (page 5) had been the first evidences of ancient human beings and that the potassium-argon method of determining the age had then been known. Scientists would have known that bipedal humans were present 3.6 million years ago. If the first fossil bones had been the Hadar Ethiopian discoveries, it would have been clear that the bipeds had primitive teeth and skulls and very small brains. Note that the discoveries lead to the same conclusion the South African finds do; but the lack of footprints led to long controversies over the mode of locomotion, and complex cave deposits caused uncertainty over associations.

Recently major discoveries of very primitive members of the genus *Australopithecus* have been found in the Afar region of Ethiopia by Donald Johanson, Maurice Taieb, and co-workers. These include numerous individuals and the most complete skeleton found so far. The remains date at approximately 3 million years, and both Johanson and Tim White believe that they represent a new species, with teeth closely similar to those found at Laetoli.[15] These finds have not yet been fully described, but some hominids have larger canine teeth than previously known hominids. The arms of the skeleton are relatively long, which are apelike features, and as such are further narrowing the gap between ape and human. But the pelvis from Afar is similar to that from Sterkfontein, South Africa, showing that these forms were bipedal and that the footprints at Laetoli were made by such creatures. The brain size of one of the Afar skulls has a capacity of less than 400 cc, the smallest of any hominid found so far.

The uncertain political situation makes more excavation in Ethiopia impossible at the present time, but we can confidently expect that the rich Ethiopian deposits will yield many, many more fossils. Only politics and money now slow the discovery of the fossils that will give us a far richer understanding of human evolution.

East of Lake Turkana

Over a period of several years Richard Leakey and numerous colleagues have made a series of remarkable discoveries in the rich fossil beds east of Lake Turkana.[16] Numerous skulls and many other parts of skeletons of hominids have been found. There has been controversy over the dating of the deposits, but 2 to 1

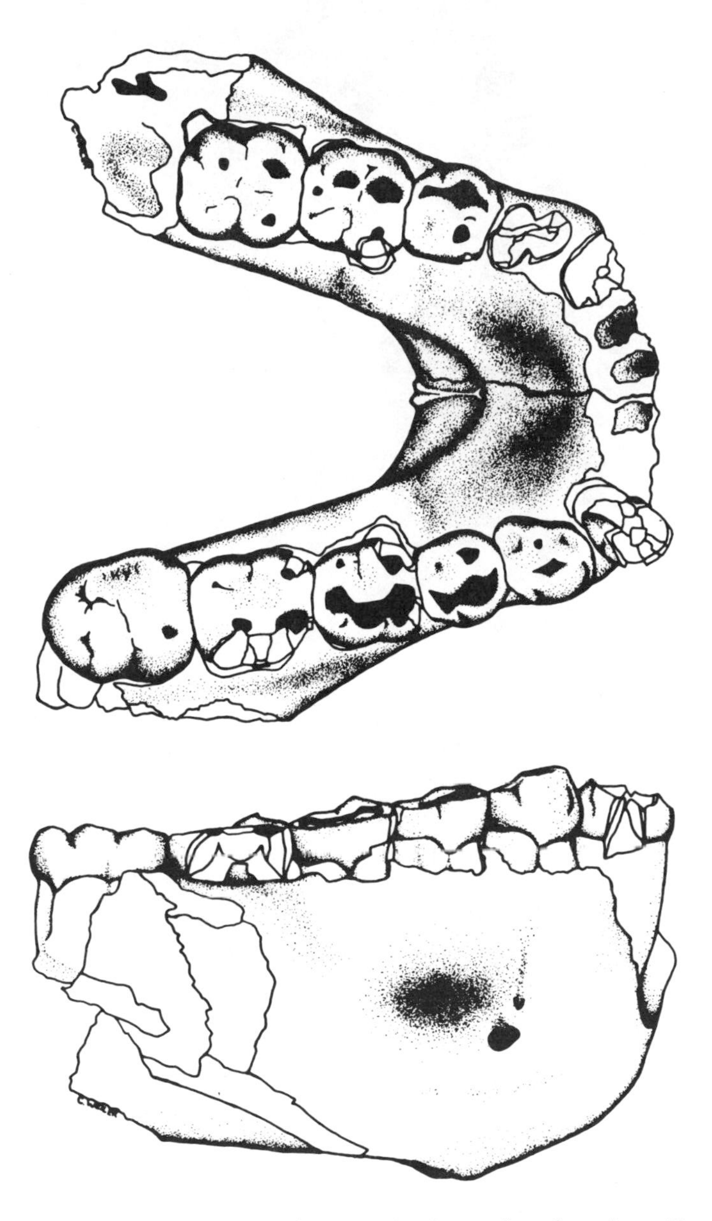

Figure 4.7 Two views of a primitive lower jaw from Laetoli; it is very similar to the much younger ones found at Afar in Ethiopia.

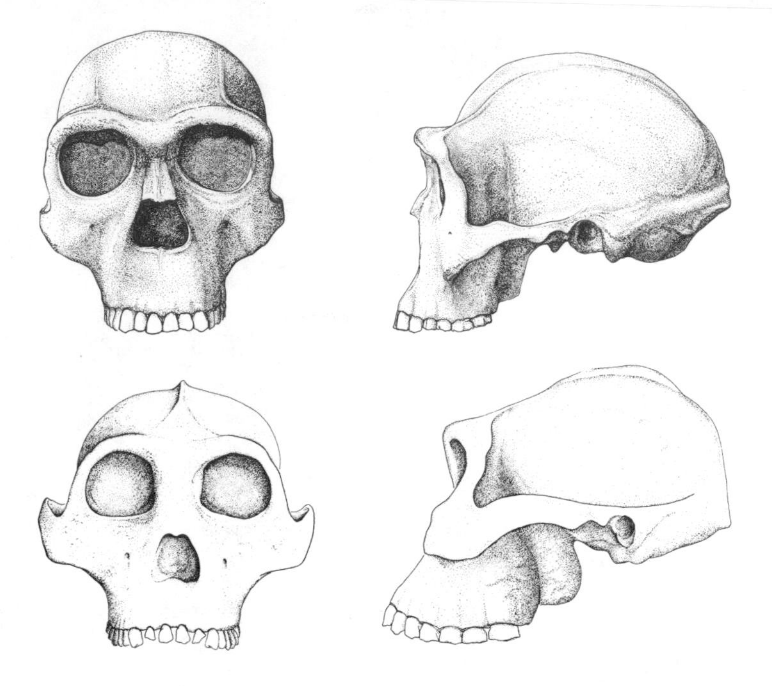

Figure 4.8 *Homo erectus* (top), and a robust australopithecine from Koobi Fora (east of Lake Turkana). These are contemporary and clearly show that there were at least two species of fossil humans in East Africa for at least a million years.

million years is a useful estimate. The skulls are of forms transitional between the earlier Afar fossils *(Australopithecus)* and excellently preserved specimens of *Homo erectus*. The robust kind of australopithecines are present and contemporary with *Homo erectus*. The evolutionary order of *Australopithecus* (3 million), *Homo habilis* (2 to 1.5 million), and *Homo erectus* (1.5 to 0.2 million) is confirmed and greatly enriched by the quantity and quality of the East Turkana discoveries.[17] Not only is Richard Leakey the most successful of the human fossil hunters, but he has also been most generous in allowing numerous scientists see the original fossils —a very different situation from that time in the past when many fossils (including Piltdown) were locked

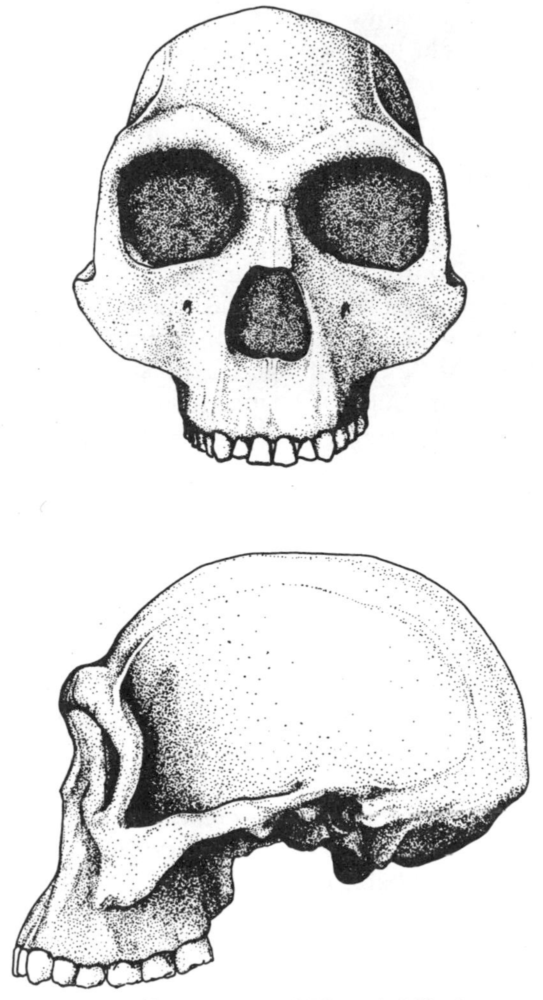

Figure 4.9 A small specimen of *Homo habilis,* humans intermediate between *Australopithecus* and *Homo erectus*. The transitional forms may have lasted for some 500,000 years between 2 and 1.5 million years ago. This remarkably preserved specimen is from Koobi Fora.

away and few were allowed to view them. (See Figures 4.7–4.9.)

One might think that the desire to find human ancestors would have led to accepting the fossils. The previously formed climate of opinion, however, led to the conclusion that the Taung skull was that of an ape, that the first pelvis did not go with the skull, that nothing with so small a brain could have made tools. It took more than twenty years of discoveries and controversy before the "obvious" conclusions were reached. For thirty years more the "obvious" conclusions were resisted by attempts to show either that the human features of the australopithecines did not exist or that all the finds belonged in a single species. It would be fascinating if a science historian would make a study of the recent controversies; it might help us understand why the fossils were, to use Le Gros Clark's phrase, "Bones of Contention."

Sterkfontein: Crucial New Light on Hominid Evolution (A Letter from Phillip V. Tobias)

The Sterkfontein limestone cave lies on a low hill not far from Johannesburg in the Southern Transvaal. Here the late Robert Broom found the first adult *Australopithecus* skull in 1936. Many more specimens of the lightly built, small-brained creature, *Australopithecus africanus,* were found before and after World War II, so that Sterkfontein became, and still remains, the world's richest single source of fossils of this species of early hominid.

In 1956, another part of the Sterkfontein cave yielded stone implements similar in form to those described by Mary Leakey from the Olduvai Gorge in northern Tanzania. The problem now arose: Who was the manufacturer of the Sterkfontein stone tools? Was it *Australopithecus africanus,* the only hominid whose bones had thus far been found in that cave deposit? Or was it a more advanced kind of hominid, such as *Homo habilis,* the probable maker of the Olduvai implements?

For 20 years the authorship of the Sterkfontein stone tools remained uncertain. A few fragments of hominid skull and teeth came to light from the part of the cave deposit that contained the stone tools, but they were so scanty that opinions differed as to what species they represented. It was puzzling, too, that the part of the cave that had yielded the rich stockpile of *A. africanus* fossils did not contain any stone tools.

A new excavation was started at Sterkfontein late in 1966, one day after the 100th anniversary of the birth of Robert Broom. One of our objectives was to clear up the stratigraphy of the cave, that is, the sequence of strata or layers in the cave filling. Also, we wanted to find out more about the dating of the various parts of the cave deposit. We hoped to ascertain the ecological surroundings of the hominids as formed by the animals and plants contemporary with them. And, of course, we wished to solve the problem of the nature of the Sterkfontein toolmaker.

Our new dig has continued for 48 weeks of every year since 1967. We have discovered that the cave filling comprises a series of strata, with interruptions separating consecutive layers. In all, there are six of these layers or *members*, and they are numbered from the bottom up [see Figure 4.10]. Thus, Member 1 is the oldest and deepest layer in the cave and Member 6 the most recent. The total depth of deposit from the bottom of Member 1 to the top of Member 6 is at least 30 meters, though it may have been greater; weathering has removed some of the cave filling.

We know now that all of the fossils of the species *Australopithecus africanus* have come from one layer, Member 4, which is far from being the oldest part of the cave deposit. That layer contains no foreign stones, that is, stones of a different material from the sort found in the walls and roof of the cave. There are also no stone implements in Member 4. On the other hand, we have numerous fossilized bones of animal species that were contemporaries of the hominid. These in-

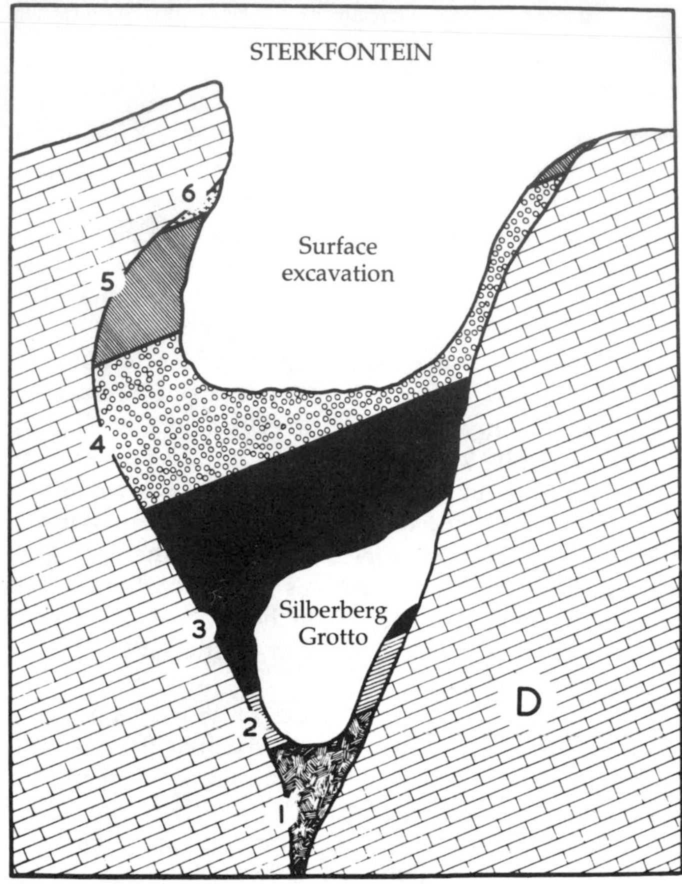

Figure 4.10 A schematic north-south section through the Sterkfontein cave deposit, showing the positions of Members 4 and 5 from which hominid fossils have been recovered. Although Member 5 lies immediately on top of Member 4, the interface between them shows that there was an appreciable lapse of time between the preserved top of 4 and the beginnings of deposition of 5. This passage of time is confirmed by the very different nature of the contents of the two Members. We may summarize these differences as follows. *Member 4:* very archaic species of animals; no foreign stone; no stone implements; *Australopithecus africanus* the only hominid identified; rich bush and tree cover; date about 3.0–2.5 million years before present. *Member 5:* more modern animals; presence of imported foreign stone; presence of stone implements; early *Homo—Homo habilis* the only hominid identified; open grassland; date about 2.0–1.5 million years before present.

clude extinct pigs, elephants, antelopes, baboons, hyenas, rodents, and others. These animals belong mainly to the species that inhabit somewhat moist regions with good vegetation cover; some fossil pollen from this stratum supports this. Unfortunately, dolomitic limestone cave deposits like Sterkfontein do not contain any materials suitable for radioactive dating techniques. In East Africa, however, the remains of ancient times are found in highly volcanic belts along the Great Rift Valley. The lava and other volcanic materials are suitable for radioactive dating. Hence we know a great deal about the dates of the extinct animals that lived alongside the hominid in Tanzania, Kenya, and Ethiopia. The animal fossils of Sterkfontein are most similar to the East African species that lived between about 3.0 and 2.5 million years ago. It is reasonable to infer that the Transvaal species from Sterkfontein Member 4 lived at approximately the same period.

Thus, we have been able to infer a date of 3.0–2.5 million years before present for Member 4 at Sterkfontein. This receives support from the fact that the Member 4 hominid *(A. africanus)* itself is extremely similar to the East African hominid that lived at about the same time, as found at Omo and Hadar in Ethiopia.

Our new excavation has shown that the stone implements of Sterkfontein emanate from Member 5: this is the layer above Member 4, and it is therefore more recent. Moreover, there seems to be a significant time lapse between the top of Member 4 and the bottom of Member 5, when erosion was occurring within the cave. Member 5 contains foreign stones and tools, a less archaic fauna, and species that indicate that open grassland was the predominant environment. So climatic and faunal conditions had changed between the time of Member 4 and the time of Member 5. What can we say of the hominid? The fragments which had earlier come to light from this layer were difficult to identify. Some thought they were *A. africanus,* others that they were early *Homo,* like *H. habilis* of Olduvai. Then in 1976, a dramatic discovery was made in Member 5: the skull and teeth of a hominid were found, 40 years to the day after Dr. Broom's first visit to Sterkfontein. This fossil (Figure 4.11) is quite different from *A. africanus,* and is the same kind of skull as that of early *Homo* in East Africa. It is considered to represent the Transvaal population of *Homo habilis.* There is little doubt now that the toolmaker of Sterkfontein was not *Australopithecus* at all but, as in East Africa, *Homo habilis.*

The evidence of the animal bones and stone tools in Member 5 points to this layer being 2.0 million years old, or even a little less. This shows how much time passed between Members 4 and 5 at Sterkfontein.

Makapansgat is another picturesque cave in the Transvaal with very ancient fauna, including *A. africanus.* From that site there is additional evidence on the dating provided by the measurement of the sequence

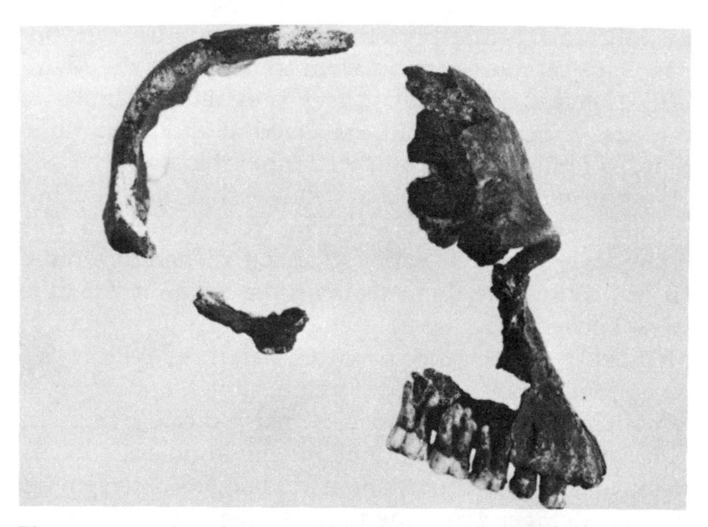

Figure 4.11 A side view of partially assembled cranium of *Homo habilis* from Member 5 of the Sterkfontein cave deposit. Discovered in August 1976, this skull provided the solution to the problem as to who was the maker of the Sterkfontein stone tools.

of normal and reversed magnetic polarity preserved in the consecutive layers of the cave filling. This *palaeomagnetic* sequence at Makapansgat shows that *A. africanus* from that site lived just about 3.0 million years ago, almost exactly the same date as the closely related "Lucy" and other *Australopithecus* fossils from Hadar in northeastern Ethiopia.

So the newest evidence strongly supports the idea that, both in South and East Africa, the ancestor of *Homo sapiens,* late in the Pliocene Period, was *Australopithecus africanus.* Later, just as the Pliocene was giving way to the Pleistocene, *A. africanus* branched into two lines of evolutionary descendants. One gave rise to sturdy, even bigger-toothed hominids known in the Transvaal as *A. robustus* and in East Africa as *A. boisei,* who died out after another million or so years, about a million years ago. The other line was of bigger-brained, smaller-toothed, toolmaking and tool-dependent hominids that eventually become *Homo sapiens,* the modern species of humankind. At least two crucial stages of this sequence are enshrined in the Sterkfontein cave, one in Member 4 (before the splitting of the hominid line) and one in Member 5 (after the branching). The earlier Members 1, 2, and 3 have yet to give up their secrets.

19th March 1979 — P. V. Tobias

5

BECOMING HUMAN

Figure 5.1 shows the relations of human beings to the apes. Clearly, almost all of primate evolution occurred before there were any human beings. The apes evolved millions of years before the australopithecines. Judging from contemporary apes, apes live in forests —either in rain forests or in woodlands close to the rain forests. Even the earliest human beings now known lived in much drier country, in open forests and savannahs; this kind of country is greatly varied locally, affording opportunities for a number of adaptations. The ancestral apes might have been less restricted in their habitat than contemporary ones, but it is a striking fact that there are no bones of apes in any of the australopithecine deposits. Thousands upon thousands

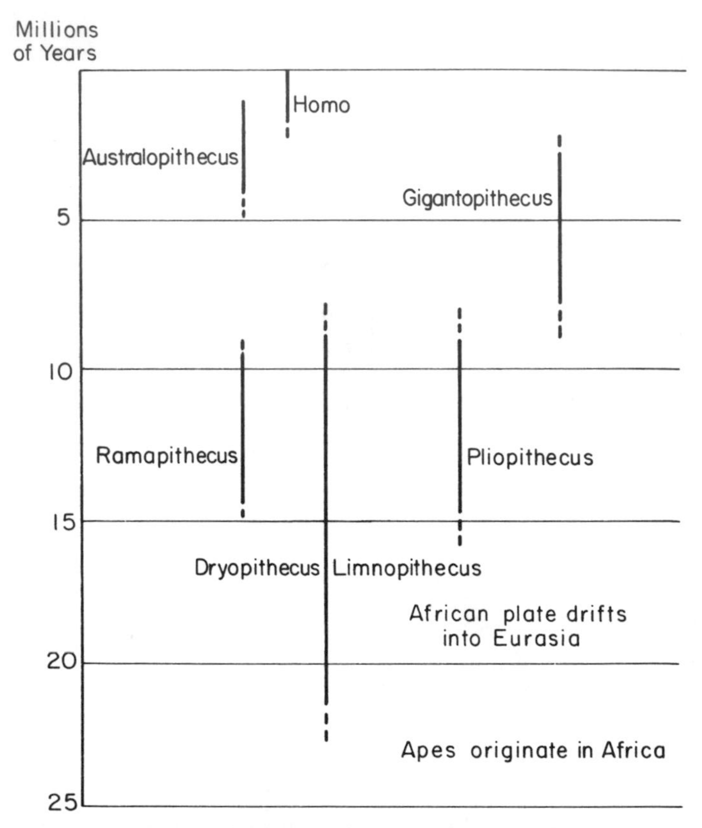

Figure 5.1 A possible view of the major kinds of fossil apes. Note that the easiest way to reconcile the paleontological and biochemical evidences is to view *Ramapithecus* as an ape, ancestral to both African apes and human beings.

of bones have been collected in a huge area extending from South Africa to Ethiopia, and hundreds of monkey bones have been found but not a single ape bone. The evidence strongly suggests that the new human way of life opened huge areas the apes had not been able to occupy.

It was a way of life that evolved. But all that remained of the past, except for the DNA transmitted to all descendants, were the bones. The bones blasted from ancient cave deposits or dug from the beaches of the past are the only visible evidence of what the hominids had been. Science had to reconstruct the populations spreading out across the drylands of Africa and perhaps farther on.

How could a human possibly have been derived from apes—apes that lived in a limited forest territory; that were primarily vegetarians and gatherers; that foraged for themselves and ate what they found where they found it; that were essentially nomads without a fixed base or home; that were dominated by the big males, whose young were not taught but had to learn by watching and in play? How did the intelligent and inquisitive apes turn into hunters, meat eaters, world travelers, and cooperative individuals? How did apes become human (Figure 5.2)?

Achieving a Unique Way of Life

Other questions had to be answered about this long stretch of prehistory. Again, and most particularly, why did only one group and no other make the

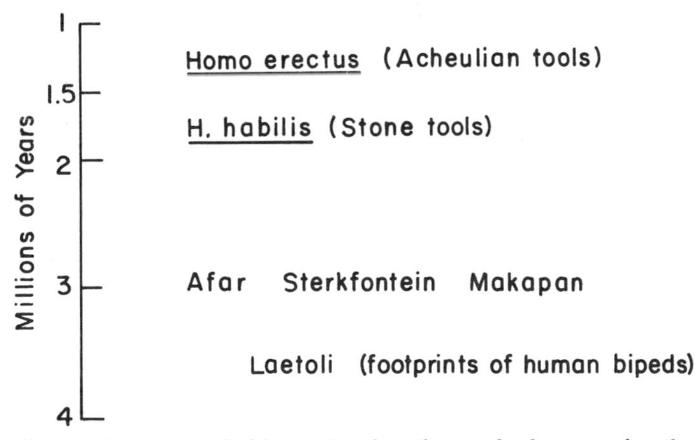

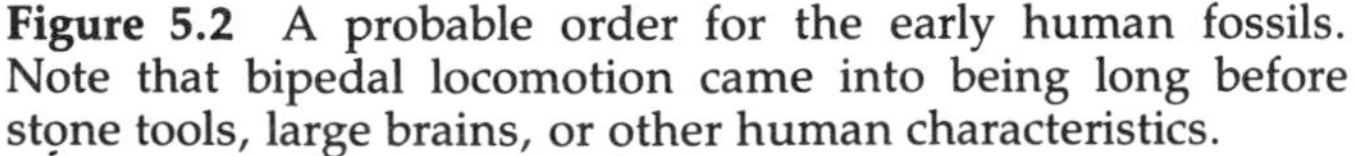
Figure 5.2 A probable order for the early human fossils. Note that bipedal locomotion came into being long before stone tools, large brains, or other human characteristics.

transition? How did one group that had broken away from the other apes become human? From the australopithecines on, certainly, the answers had to be sought in the unique way of life that was evolved and in how this changed or did not change the early humans physically and biologically.

The fossils, of course, supplied one clue to the biggest change ever made by any living group. Studying the living populations of apes yielded another part of the answer, for some had made little change in the ancestral way of life from which the hominids had departed. The new biochemical studies also were drawn upon again for information about what had changed and what had not.

If what evolved was *successful behavior,* understanding of the evolution of the species still could be achieved only by the reconstruction of the behaviors of past populations. But the more well-dated fossils that are available and the more that is known of the behavior of living primates, the more reliable the reconstruction of the past can be.

Well before the hominids moved out of the forest they may have used tools. At what point the ape's waving or throwing of a branch or a stone became a regular or an aimed wielding of these weapons no one can yet say. No timed evidence shows when the chimpanzee's stripping of a twig to make a tool for termiting turned into using a chunk of rock to pry or to break open a nut. Experiment confirmed the huge advantage that a stone tool gives to its user. Held in the hand, it can be used for pounding, digging, or scraping. Flesh and bone can be cut with a flaked chip, and what would be a mild blow with the fist becomes lethal with a rock in the hand. If monkeys used a stick or stone to dig out roots, they could double their food supply.[1] Exact data also are lacking on when the great apes' carrying of a stick or piece of fruit in one hand during knuckle walking changed into an upright walk or shuffle that permitted sticks or fruit to be carried in both hands. It was clear, nevertheless, that bipedalism freed the hands for a wide variety of skills. It made possible the human way of life.

In many populations of apes over some millions of years minimal tool use was present. In some of these populations the carrying of tools and the products of tool use became so important that selection favored those groups of apes in which bipedal locomotion was most efficient. Bipedalism permitted the evolution of skillful, practiced tool use, which in turn became more effective as the locomotor pattern evolved responding to the new demands. Thus, locomotion and tool use were both cause and effect of each other.

Tool use did not result from a preceding bipedal adaptation nor from a discovery that happened once. Tool use probably developed through history repeating itself countless times in ape populations of several species over millions of years. By the time our ancestors had taken to life on the wide dry savannahs of South Africa or in the bush country of East Africa they had become tool users and toolmakers.

The evidence of this new way of life is twofold —the tools themselves and the reduction of the canine teeth. All the australopithecines found so far, in South, East, and Northeast Africa, have small canine teeth. Without the canines and other defenses, such as speed or ruse, they would not have survived very long on the savannah. They would have been easy prey for carnivores, and the small canines in themselves are proof of reliance on tools and weapons.

By 2 million years ago—though the date is an estimate—the hominids of South Africa had learned to make and use pebble tools. They carried pieces of quartz and chert from the beaches of nearby rivers to their caves. Nearly three fourths of the pebbles were not worked at all. If there had been any similar stones in the campsites where they were found, they could not have been identified as tools.

Since chimpanzees use stones for a variety of purposes, a bipedal human would have used them with even more versatility. Some stones would have broken from use, and the usefulness of the sharp fragments would have been a simple discovery. The archaeological evidence, however, suggests that our ancestors were bipeds for at least 2 million years before stone-tool making became common.

To skin a large animal with the teeth and fingernails alone is nearly impossible, as Louis Leakey once demonstrated. When he knocked a few chips from a rock and used the sharp edge to strip away the skin, he had the meat in a few minutes. Some pebbles, though, were deliberately made into tools. When a rock was

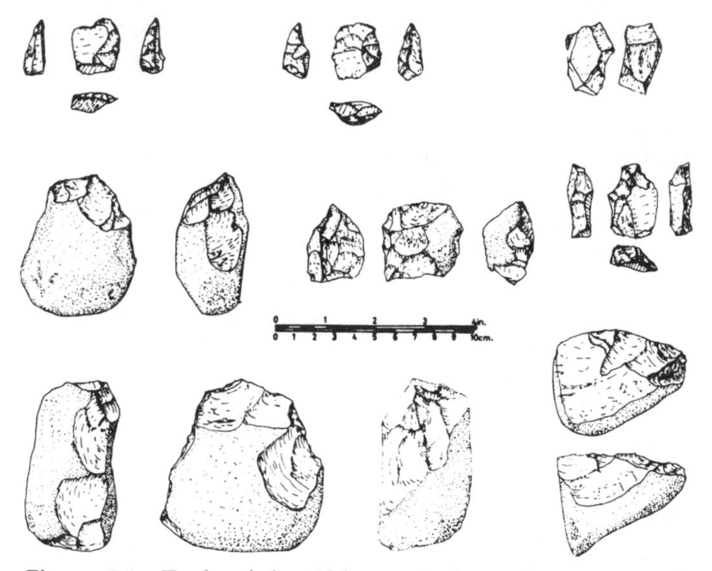

Figure 5.3 Tools of the Oldowan Industry from Bed I, Olduvai Gorge (above), which have been dated to 1.75 million years ago, are contrasted with the Upper Acheulian tools from Kalambo Falls (opposite page), which have been dated to 60,000 years before the present and earlier. The primitive Oldowan tools, which are made from quartz and lava, contrast with the specialized tools of the Acheulian period, which are made from chert (the small size) and from quartzite (the larger).

held with one hand and struck with another, a chip was knocked off one side and then the other, producing an effective cutting edge.

J. Desmond Clark gives a course in prehistory at the University of California to afford graduate anthropology students an insight into the life of early humans. With the guidance of their instructor, the students learn to make good pebble tools in a day. However, even in a full semester of work, no student has mastered the making of the fine stone tools that were made much later in our history. (See Figure 5.3.)

At Olduvai the discovery of the living floor or campsite of the *Zinj* band and its dating made it certain that the hominids were making stone tools at least 1.75 million years ago. Both numerous pebble tools and also

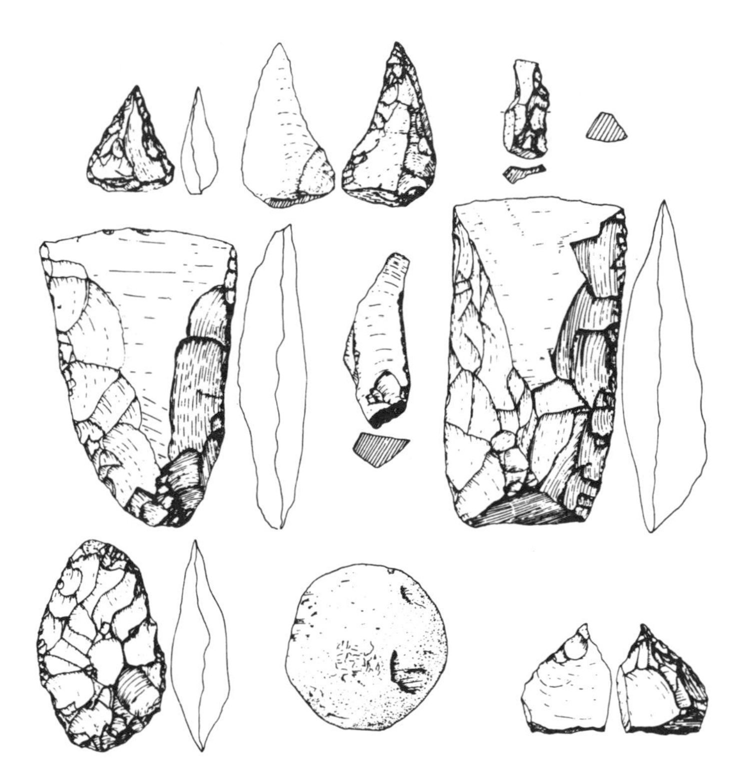

the hammerstone that had been used were found. Some chips of stone that had been struck off also showed signs of use. The sharp-edged chips were excellent for scraping or for digging the marrow from a bone. In the early 1970s a comparable living floor, found by Richard Leakey at Lake Turkana, was dated at about 1.8 million years ago.

The hand that made the sharp-edged pebble tools was, as the discovery of the bones proved, a hand between an ape's and a human's. Though relatively small, the proportions of the digits and the palm were much the same as in humans. The tips of the terminal bones of the fingers were quite wide, and the fingertips must have been broad. The fingers were somewhat curved.

John Napier, an anatomist who studied the Olduvai hand and the evolution of the hand, pointed out that the Olduvai hand would have been capable of a tremendously strong power grip, perhaps stronger than that of modern humans.[2] (In the power grip, a tool such as a screwdriver is held in the palm with the fingers flexed around it and the thumb applying counterpressure. Essentially a whole-hand grip, it is used when strength is needed. The other basic grip, the precision, is used when accuracy and delicacy of touch are required: a tool such as a pencil is held between the tips of one or more fingers and the fully opposed thumb—it is basically a finger grip.)

Napier commented:

> The inception of toolmaking has hitherto been regarded as the milestone that marked the emergence of the genus *Homo*. It has been assumed that the development was a sudden event, happening as it were, overnight. It is now becoming clear that this important cultural phase in evolution had its inception at a much earlier stage in the biological evolution of man and that it existed for a much longer period of time and that it was set in motion by a much less advanced hominid and a

much less specialized hand than has previously been believed.[3]

Making and using tools therefore was part of the new way of life the australopithecines were taking up on the savannah. They used tools instead of teeth in fighting, as well as when hunting and preparing food. Yet the stone tools found with the hominids must have been only a fraction of their tools. Sticks and bones must have been used as well. The chimpanzees, as the Goodall work showed, made termiting sticks and sponges. However, these acts in the forest apparently were not sufficiently important to alter selection pressures in their favor. In the world of the forest, capturing termites and recovering water from a pocket in a branch did not determine survival or death. Only when populations moved out onto the drylands was using and making tools vital.

Hunting, Meat Eating, and Population

On the savannah there were few fruit trees and not as many of the succulent herbs in which the forest abounded. But small game was profuse. The great apes had always eaten meat with great eagerness and relish when they happened to kill a monkey or a young antelope. Some of the meat might be shared, but the apes were primarily vegetarians and the hunting was occasional and often seemed to occur by chance.

With the hominids the old pattern was reversed. The bones around the sites where they camped or lived showed that they ate whatever animals they could capture and kill. At the same time, they did not abandon the old vegetarian diet. Life, however, began to center around meat and hunting.[4]

Some have suggested that the australopithecines may have started their meat eating as scavengers, seizing what they could after the lions had made the kills. The studies of the modern great apes throw this theory into question. Both chimpanzees and gorillas ignore a carcass lying on the ground; only when they make a kill do they eat meat. Nothing indicates that the hominids with their simple stone tools could have driven a lion away from his kill or could even have routed the hyenas or vultures moving in after the lions had left. Scavenging would have brought the australopithecines into direct conflict with large animals they could not beat (see Figure 5.4).

The australopithecines, then, had to hunt for their food in the savannah into which they had moved. With stone in hand an australopithecine was well equipped to kill a small or young animal. The stone could crack open a small skull as an empty fist never could. Or a stone thrown, either underhand or overhand, might hit and cripple an animal, making its capture much easier.

Studies of the baboons (see Figure 5.5) have shown how another animal taking up life in the same dry open lands makes similar adaptations. The baboons, among the few monkeys that have moved to

Figure 5.4 Lioness with wildebeest she has killed. Predators protect their kills and there is little left when they have finished.

Figure 5.5 Observing baboons in the Nairobi Park. The animals are used to humans and very close observation is possible.

the ground, also became partial meat eaters and hunters. They made no progress in learning to use tools, but the big powerful males learned to hunt in groups. They surround their prey and block its escape. With their large canines (see Figure 5.6) they can dispatch almost any small animal.

The presence of a little band at Olduvai with the game they had eaten indicated that the hominids may have also turned to group hunting. When an animal was killed, the hunters may have carried or dragged it back to their campsite. In contrast, the baboons neither carry nor store food. Carrying is so awkward for quadrupeds that they transport nothing in the hands for more than a few yards, though they may occasionally carry food or some other object in the mouth or under an arm. One observer once saw a baboon proceeding by tossing an object ahead, picking it up, tossing it again, and so on.

In Sterkfontein the presence of stones that had been carried for miles from a river gravel is evidence for bipedalism, and the structure of the australopithe-

Figure 5.6 Baboon yawning. Note the size of the teeth and that individual teeth can be seen, enabling researchers to estimate the age of the animal.

cine pelvis supports this inference. It is hard to believe that animals that carried stones did not also carry food and thus initiate an important stage in freeing humans from the necessity of eating their food where they found it.

There were very few human beings until after the origin of agriculture. There now exist enough carefully examined excavations so that the number of hominids (*Australopithecus* plus *Homo*) may be compared with that of baboons. The baboon fossils are common, possibly 50 to 100 times as many as hominid fossils. At Omo there were about as many hominids as there were carnivores.[5]

If these early humans were killing and eating other animals, they were not doing it the way carnivores do. The canine teeth of carnivores are large in both sexes and are essential for killing and tearing game. Even if we take the pygmy chimpanzee as a standard, human incisors and canines are smaller and the sexual differentiation is mostly lost. The specimens from Laetoli and Afar show primitive features in both the canines and the first lower premolars, signifying that the last of the reduction took place after the evolution of bipedal locomotion. The molar teeth of hominids, however, were larger than those in the apes. An adaptation that explains these differences is the hominids' tool using and meat eating. When chimpanzees eat meat, they chew it, sometimes for hours. If the ancestral form was a fruit eater, like the chimpanzee, and meat eating became important, selection would have been for larger molar teeth. In carnivores teeth adapted for chopping to cut the meat, and large pieces may be swallowed; in hominids meat must be chewed, and digestion starts in the mouth.

Meat eating, judging by the chimpanzee or the baboon, means eating tough meat, skin, and bits of bone—and eating most of an animal, not only selected parts. At the rate of chewing reported for chimpanzees, an animal could not eat enough to survive. Another way of considering the situation is the question of why there were so many more baboons than hominids if our ancestors could eat what baboons do.

To summarize, the model presented here is that the ancestral apes were knuckle-walking fruit eaters and that the success of object using (for both defense and killing, within the same species and with others) led to selection for bipedal locomotion. There is very little fruit in the savannah. Life is more dangerous. A tool-using biped adapted to the problems of the savannah by hunting and eating meat.

This scheme accounts for why there were so few hominids, why they became bipedal, and why they had large molar teeth. If there were large sex differences in body size, as suggested by the specimens from Afar, such differences would support the notion that fighting and hunting had shifted from teeth to tools.

The purpose of a scheme such as this is not to find truth but to see evolution as progress toward more and more effective adaptations. The units of evolutionary change were functional patterns, and such patterns are interrelated. For example, we now know that some apes may be territorial; they fight neighboring groups and may fight strange individuals. Sticks and rocks may be used both in threatening displays and in actual fighting. The evolutionary importance of objects in intergroup competition, even in the earliest stages of human evolution, is something that could not have been anticipated before the fieldwork of the last few years. The peaceful apes pictured in the work of the 1960s have been replaced by animals that may be aggressive, defend territory, and even eat their own kind.[6]

Water

The ability to carry water would greatly increase mobility. Terrestrial monkeys, such as baboons, must stay within about two miles of water. When the australopithecines went to the river to gather stones and possibly to hunt animals along the banks, they did not seem to gather shellfish or fish. The consumption of shellfish leaves huge middens that tend to last indefinitely when a site is buried, and none has been found.

Except for drinking at the edge of a stream, water meant danger. To learn to swim, the bipedal hominid would have had to master a difficult skill. Unlike the quadrupeds, the bipeds cannot swim naturally. The quadrupedal monkey, like a dog, can keep afloat with its normal running movements and thus does not have to learn a new motor habit to swim. This began to change with the apes. Their normal patterns of movement will not keep them above the water's surface, and George Schaller saw that gorillas will not cross even a narrow stream if they have to enter the water. Water must have also been a barrier for the hominids. Though they went to the edge, they stayed out and did not extend their hunting into the stream.

Farther Afield

Hunting began to take the hominids farther afield. Most of their primate ancestors lived their entire lives in a range of four or five square miles. Monkeys have excellent vision, and from the treetops they undoubtedly can see fruit trees and promising land much beyond their own range. Whatever the lure may be, they do not go. Attempts have been made to drive baboons outside their range, but when the animals approach the boundaries they become uneasy and excited. If driven farther, they loop back to their familiar land. The hominids with their stones in hand ventured farther to get enough meat to satisfy their needs. Not until agriculture came in several million years later could humans live on a small piece of ground.

How much territory is necessary for a hunter is demonstrated by the Bushmen of Africa. One area of 600 square miles with eleven water holes supports 240 hunters, while in a nearby preserve about 250 baboons live in a few square miles around one water hole. In Australia bands of about 35 aborigine hunter-gatherers occupied home ranges of 150 to 750 square miles. Constant hunting, by humans or animals, rapidly depletes the game in a small area. The 40 square miles of Nairobi Park, for example, support an average population of only 14 lions as compared with 400 baboons.

The wider areas beckoning the hominids also gave them access to more seasonal foods and to more potentially useful materials, important for increasing tool use. The pebble tools showed that the hominids went a considerable distance to find rocks of the right size, form, and material. They may also have gone to more than one stream to make their selection. Wood also had to be found. When wood became something more than a tree to climb, hardness and texture counted. If a stick were to be used, it had to be stiff and strong enough not to break at the first impact.

Availability of materials is critical to the tool user, and early humans must have had a very different interest in their environment from that of monkeys and apes. The presence of tools in the archaeological record is an indication not only of technical progress but also of interest in inanimate objects and in a much larger part of the environment than is the case with nonhuman primates.

In a small area the population must be carried on local resources, and natural selection favors biology and behavior that efficiently utilize these limited opportunities. In a wider area natural selection favors the knowledge that enables a group to utilize seasonal and occasional food sources. Interest in a large area is human. When the australopithecines could walk twenty miles in a straight line, they might be said to be human and unique.

Early Groups

The hunters who walked twenty miles in a straight line (it may have been a zigzag) in pursuit of some quarry were members of a group. Every campsite testifies to the presence of several individuals. Like most primates, the australopithecines then were social, group-living creatures. Moreover, the group itself was a survival mechanism.

Studies of the contemporary primates as well as the archaeological record supply some strong indications of the size of the australopithecine bands. Lee and DeVore believed that bands of hunters were small.[7] Kenneth Oakley suggests a range of ten to two hundred. Such group sizes are common in the nonhuman primates.

The baboons, the other specialized savannah dwellers, most frequently live in troops of thirty to fifty. Such a group might have five to ten adult males, ten to twenty adult females, and an equal number of juveniles. One typical group near Victoria Falls was made up of five adult males, twelve adult females, fourteen juveniles, and two infants. At the other end of the scale, the size of bands of primitive human hunters has been estimated to average from twenty to fifty. It appears that the size of the band did not increase greatly and could not until after agriculture. The human way of life does not necessarily result in an increase in the size of the local group.

Anthropologists also put the comparison in terms of population density. In Nairobi Park there are about ten baboons per square mile, and in other parts of Africa the density is believed to be greater. But preagricultural humans, the studies indicate, would need five to ten square miles per person.

In other words, troops of baboons may exceed in size the bands of primitive hunters, and the density of baboons may reach a hundred times that attributed to our ancestors. Even today there are more baboons than Masai in the Masai Reserve.

In the little australopithecine bands the ape way of life changed into a new human way. Membership in the group directly influenced survival and reproduction. The group offered protection against predators, help in finding food and water, and a way to cope with injury and illness. It also facilitated the production of young and their care and training.

The base was established much earlier. Without its group a lone primate had a very poor chance of surviving membership in the group meant life or death. Once a band began living on the ground, the much greater danger from predators alone (see Figure 5.7) would be sufficient reason for a strictly organized social system.

Observers often watched primate groups moving through the forest. Any animal unable to keep up with the group soon disappeared, presumably captured and eaten by predators. The young and the sick or injured were in particular danger if they became separated from the group. When the troop moved out on the

Figure 5.7 Even primates as large as baboons must seek safety from large carnivores in trees. On the open savannah, protection from predators is a major problem for contemporary primates, as it must have been for early humans. The baboons in this photograph taken at Nairobi National Park, Kenya, have climbed a thorn tree to avoid the lioness.

daily round, all members had to move with it or be deserted. Sick and wounded animals often made great efforts to keep up with the troop but finally fell behind. In one case, at least three of these were killed. Among wild primates injuries are common, and animals so sick that they can be spotted by a relatively distant human observer are frequent. For a wild primate a fatal sickness is one that separates it from the troop. An infant separated from its mother and the troop had almost no chance of surviving. Only those who succeeded in staying with the troop lived to pass on their skills to offspring. When an attack was made on a troop of apes, the males moved out to meet it. The females and young took shelter in the trees if any were available. The troop also guarded against surprise (see Figure 5.8).

There is no indication that the role of the male as the protector and the fighter changed when the descendants of the apes began fighting with stone tools instead of teeth. It was also the male who went hunting and first pushed on beyond the psychological boundaries that always had halted all other primates.

The female australopithecines became gatherers. They apparently picked fruits and roots and brought them back to the camp or cave for the others to share. Thus another human change was occurring: this had not generally happened before. Except for nursing her infant, the female ape, like the male, ate her food where she found it. As soon as the infant was weaned, it had to gather food for itself, even though it remained in the company of the mother. Only the chimpanzee will sometimes share a bit with an infant or help an infant pick a piece of fruit.

Sharing: Dominance Yields to Cooperation

The change to a hunting and carrying way of life changed this general winner-take-all (or nearly all) to a more human pattern.[8] At first, as in some hunting societies, the hunters may have eaten some game at the point of capture. However, with a stone tool they could cut away some meat to carry back to the camp. The bones found on the living floors probably represented only a small part of the animals that were killed. If the whole carcass had been dragged back, the bone middens or garbage dumps would have been much larger than they were. Whether the females had found any-

Figure 5.8 A baboon troop in march position (above) and being threatened by a leopard (opposite page).

thing equivalent to a basket for bringing back fruits and leaves has not been determined.

The new hunting was a set of ways of life. It involved a division of labor between male and female, sharing according to custom, cooperation among males, planning, knowledge of many species and large areas, and technical skill. Sharing both in the family and in a wider society is fundamental.

Human hunting as a whole social pattern has often been compared with the hunting of wolves rather than with the hunting of apes or monkeys. But this completely misses the special nature of human adaptation. Human females do not go out to hunt and then regurgitate to their young when they return. Human young do not stay in dens; they are carried by their mothers. Male wolves do not kill with tools, butcher, and share with females who have been gathering. The success of the human way dominated evolution and determined the relation of biology and culture for thousands of years.[9]

Ways of living together began to change—perhaps had to change—when hominid hunters brought

food back. Social relationships in all the primate groups always had been complex, but most groups in the past were hierarchies dominated by one or a few mature males (see Figure 5.9). Each member had its place and knew it. The new primate studies have disclosed, however, that none of the primate societies were the "primal hordes" Sigmund Freud theorized in his picture of a troop as a horde in which a single aggressive male monopolized the female and drove out his sons.

In baboon troops dominance is based on the ability to fight. But once established, the dominant animal's position generally is recognized by all others. Only occasionally does the dominant have to enforce the right submissive response by a bite or a chase. Generally a big silver-backed male gorilla has only to nudge another member of the troop to keep it in line. Most chimpanzees also make the proper crouching, half-bending turn of the back that signifies submission. The crouch is used generally when subordinates come into close proximity to a high-ranking chimpanzee. When all goes well, the dominant chimpanzee may then reach out to give a reassuring touch or even embrace. (See Figure 5.10.)

However, the highest-ranking chimpanzee in the Gombe troop seemingly won his position by learning to make a great amount of noise. He discovered how to seize an empty kerosene can and drag it behind him.

Figure 5.9 A charging display can help a male's rise to dominance.

As it bounced over the ground, the other chimps hurried to get out of the way.

Sometimes the dominance hierarchies in chimpanzees were backed up by alliances between several males. The allies would quickly go into action together if trouble arose. The observers also saw a few alliances for power between a male and female.

Figure 5.10 At top, the chimpanzee on the right crouches in submission. In the bottom photo, the chimpanzee on the left gives a reassuring touch, and the one on the right relaxes.

The dominance arrangement usually reduced intragroup fighting to a minimum. One monkey group in the laboratories of the University of California lived peaceably under the dominance of one male. When he was removed, the peaceful old order continued for two to three weeks; then the male who had stood second made his move to take over. Fights broke out, and in the next few weeks several males received severe bites. The order then was settled again, and peace returned under the new boss. Again every troop member knew its place. Dominance usually has little effect on food gathering. The normal spacing of a troop of gorillas or chimpanzees as they forage through the forest keeps the animals far enough apart for each to gather without interference from another. Rarely does a dominant animal displace another from a choice place in a fruit tree.

The gathering and immediate eating of fruit, grass, and roots led to a noncompetitive economic life. This quiet noninterference changed rapidly, however, when Goodall supplied bananas for the chimps to get. It probably changed in the same way when the use of tools and the transporting of food began. Problems of distribution were created that could not be solved by a social system under the control of the larger animals. With the killing of large animals, the problem of distribution must have become even more severe.

Although the carnivores share food, the australopithecine situation must have been much more complicated. If the troops had as many as fifty or sixty members, as some authorities have estimated, it meant that they all had to share the meat. No comparable situation exists among lions or other carnivores. The problem created was a new one. Tools, carrying, and hunting helped the early humans to survive the rigors of life on the savannah, but cooperation and sharing were also essential for survival.

With sharing, dominance very likely became less important in social control. If a dominant hominid had taken all the food, the rest of the group might have perished in hard times, and the dominant could not have survived alone or have left more offspring. Selection at last began to give the edge to cooperation. The groups that shared the meat became those that flourished and continued.

The base for such cooperation was laid within the troop by well-established relationships, most particularly by the relationship between mothers and offspring. Two continuing studies, one of Japanese macaques in Japan, and one of Indian macaques living free on the Caribbean island of Santiago, just off the coast of Puerto Rico, show how genetic relationships play a major part in setting the course and nature of social interactions. The bonds between mother and infant persisted into adult life and often provided a nucleus for other social relations.

In the Goodall study, when Flint was born to Flo, his sister Fifi was fascinated, and his two adolescent brothers, Figan and Faben, stared at him curiously and patted him occasionally. But Fifi would constantly quit her play to watch the baby and to beg her mother to let her take him. Not until the infant was about three months old did Flo let her hold him. After that she took the baby constantly, playing with him and grooming him. Flo rushed in only if she heard the infant whimper. At about six months, Flint took his first steps away from his mother. Figan and Faben then also would play with him. The mother and her three offspring formed a steady group within the troop. (See Figure 5.11.)

The relationships that continued into adult life may also have entered into dominance interactions. Donald Sade of Northwestern University saw a female rhesus monkey divert the attack of a dominant male from her adult son, and saw another adult female protect her juvenile half sister (paternity is never determinable). Goodall was watching once when Flo attacked

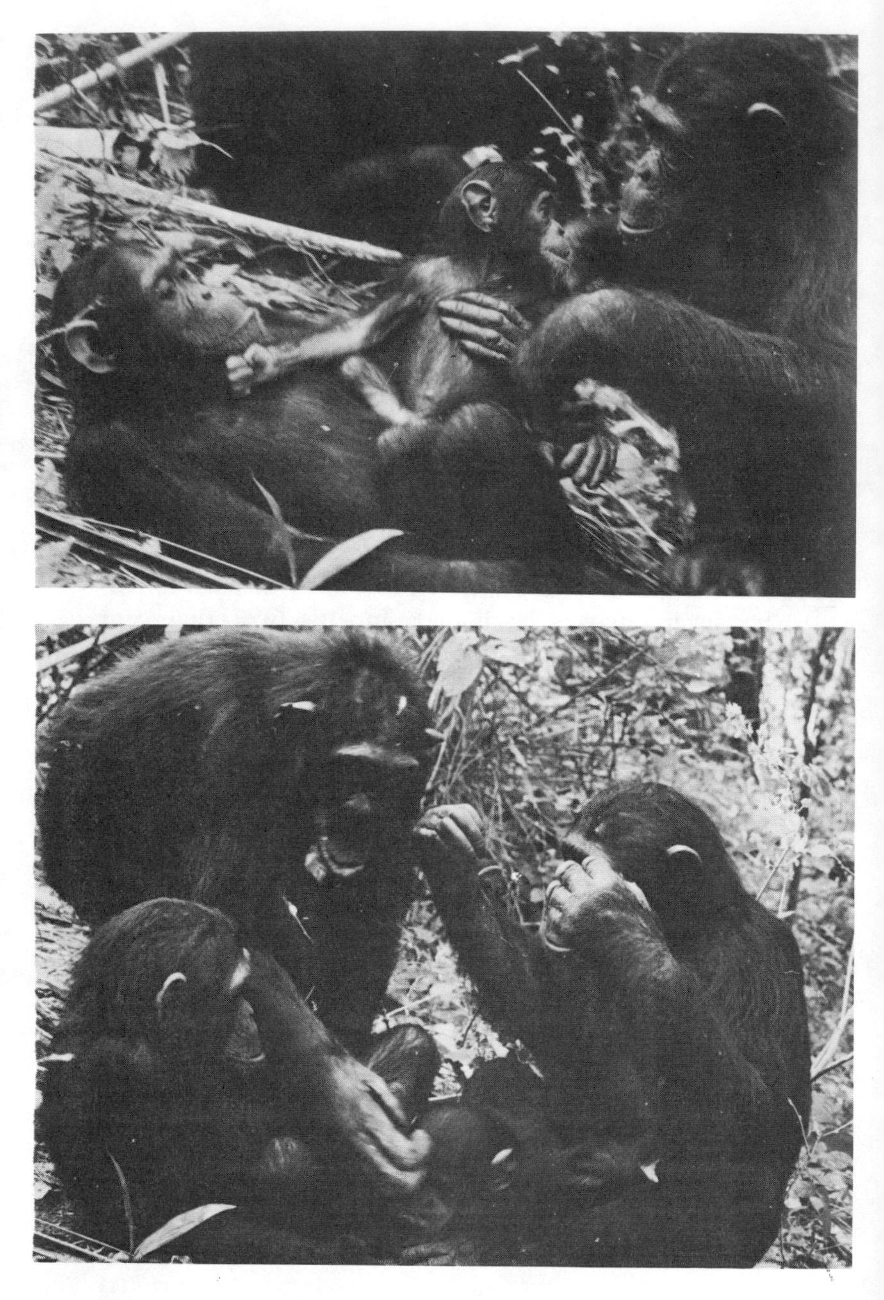

another female, perhaps without noticing that her annoyer's adolescent son, Pepe, was feeding nearby. When Pepe heard his mother's screams, he quickly came to her help, and the two together put Flo to rout.

Washburn, Jay, and Lancaster said in a paper on the subject:

> It should be stressed that there is no information leading us to believe that these animals are either recognizing genetic relationships or responding to an abstract concept of family. Rather these social relationships are determined by the necessarily close association of mother with newborn infant, which is extended through time and generations and which ramifies into close association among siblings.
>
> Because of their dramatic character, the importance of dominance and aggression has been greatly exaggerated compared to the continuing, positive affectional relations between related animals. It is expressed by their sitting and feeding together, touching and grooming.[10]

Close, affectionate relationships existed between mother and offspring, between the youngsters playing together, and in friendships that flourished between males and sometimes between males and females. The base was well laid for a human social system. When hunting and bringing home the meat required cooperation and sharing instead of the old dominance pattern, there were relationships on which sharing could be built. The little bands of man's ancestors were so constituted that social learning was not difficult.

Monkeys and apes adapt by their social life and the group provides the context of affection, protection, and stability in which learning occurs. Society is a major adaptive mechanism with many functions.

Figure 5.11 Infant chimpanzee sitting on its sister. An interested male approaches. The infant's mother approaches and the male lifts its hands, showing that he is not injuring the baby.

The Family and the Slow Maturation of Infants

The human family is a socially recognized relationship between males and females with social, economic, and sexual functions. Clearly, the economic functions and social recognition of local customs are unique to human beings. Thus, people have sought the roots of the human family in sexual attraction and behavior. First it was believed that apes were promiscuous and later that sexual behavior was limited to a period of sexual receptivity (heat, estrus, determined by hormones). It was believed that the primate social group was held together by sexual attraction that lasted throughout the year. It was not seasonal, as in the case of so many other mammals.

Human sexual behaviors were believed to be unique, both because mating was no longer limited to the brief estrus periods and because mating was performed face to face. Much has been made of the social and psychological importance of these supposed new conditions. But recently E. S. Savage-Rumbaugh has shown that pygmy chimpanzees mate throughout the cycle and use a wide variety of copulatory positions, including that of face to face.[11] Actually, both orangutan and gorilla use a wide variety of positions, and the importance of the face-to-face position seems to be a European myth rather than the biological reason for the family.

A survey of the behaviors of the apes will put the matter in perspective. In the small apes (gibbons), males and females mate for life. They live with their offspring in a small, defended territory, and they are the least sexually active of the apes. Orangutans mate at any season, and the male mates whether the female wants to or not. But orangutans are the least social of the great apes, since the males are normally solitary. In the common chimpanzee, the females are sexually receptive at all times, with highest frequencies during estrus and lowest during lactation.[12] As noted, the pygmy chimpanzee is even more human in both frequencies of mating and positions. The social group is often divided, and continual receptivity does not lead to the formation of lasting pairs. In gorillas, females are receptive only for two to four days of the month,[13] yet the gorilla group is the most cohesive of any of the great apes.

Clearly, sex is important, but equally clearly a wide variety of different patterns of behavior will result in all females becoming pregnant. Sexual behavior does not determine the form of the social systems in the apes.

The differences in the social systems may be understood by considering diet and locomotion. The large orangutans are primarily fruit eaters. Galdikas reported that orangutans would eat nothing but fruit if they could get enough.[14] Orangutans move slowly in the trees compared with the monkeys in the same area. The combination of body size, fruit-eating behavior, and locomotion determine that adult animals must be widely separated most of the time. The chimpanzee diet is over 90 percent fruit, but in contrast to the orangutan chimpanzees obtain enough fruit by knuckle walking from tree to tree. The frequently divided social group is a means by which the chimpanzees can harvest over a wide area. The gorilla's diet comes largely from ground plants. The sources of the diet are locally abundant, so the animals may remain close together. The big males protect the social group, a behavior that is necessary on the ground.

Most of the fighting is not directly the result of rivalry over a female. It is not that a male is denied access to a female but that males are more aggressive during the mating time.

From the point of view of human evolution, it is not possible to determine the time of the origin of the family. It is possible to indicate the nature of the transition. The wide variety of sexual behaviors among the apes suggests that sex as such did not cause the human family. The fundamental new factors are: sharing of food, division of labor with males hunting and females gathering, delayed maturation of young, and control of emotion. The final loss of estrus behaviors may have been very important in making an orderly social life a possibility. All these factors depend on tools and weapons. These can be found in the fossil record, and the evidence for stone tools and hunting (possibly by 2 million years ago) may be the earliest evidence that the human kind of social system based on economic reciprocity was becoming possible.

The relation of sex to the social group is made even clearer in the behavior of some monkeys. All mating may take place in one season, but the social group continues throughout the year. The functions of being social (information, protection, play mediated by the biological ease of learning to be social) are continuous, and sexual attraction is added at only one time of the year and primarily only for part of the month. Further, estrus may cause fighting among the males. In rhesus monkeys on Cayo Santiago many males are killed during the mating season.[15]

In monkeys and apes the newborn infant is quite mature by human standards. Two or three days after birth the little monkey can move about and cling to its mother. In gorillas, the slowest to mature of the apes, the mother may have to help the infant for as much as six weeks. In contrast, the human mother must carry her baby for more than a year, and for a full three years the infant can walk only short distances on its own. There has been selection for delayed maturity, in spite of the fact that the result dominates much of human life. Under past conditions human females were always taking care of the young, and although customs vary widely among various peoples, human males always help by providing economic contributions, social support, and protection. In the apes there is no comparable situation, and the young must provide for themselves as soon as they are weaned.

Bipedalism freed the woman's arms for carrying a baby, but she had to *want* to carry it. Psychiatrist David A. Hamburg, in his article, "Emotions in the Perspective of Human Evolution," said:

> The mother must find pleasure in holding the baby. She must experience some unpleasant feeling if she is deprived of the opportunity. This in fact turns out to be not so simple. It is a remarkable evolutionary achievement.
>
> There are several mother-infant transitions which probably served to strengthen the motivational-emotional bond—close bodily contact, nursing, smiling, patting and stroking, and rhythmic movement. The likelihood is that some or all of these situations strengthen the mother's motivation to care for her infant. Indeed it is reasonable to surmise that selection has favored infants whose behavior most effectively elicited caretaking emotional patterns in the mother.[16]

The biological cost of this prolongation of infancy in the ape and early hominid was high. The mother had to carry or care for the infant for three or four years. Selection could favor slow maturation only if the advantages of prolongation (of the protected learning period) were great. The added years supplied the time to learn the skills required by the new way of life, the hunting, and the wider territory.

Youth became a protected period for learning, and what was learned—through both practice and play—was exactly what was important for adult life. The young apes chasing one another and charging were practicing the behavior they would use as adults. When it came time to fight, they would know every move. Play involves an incredible amount of repetition, and this repetition is necessary for the development of the nervous system. No one would think that a person could be good at throwing a baseball without putting a tremendous amount of time into throwing a baseball. Social skills take the same kind of repetition over the years that sports and technical skills do.

The sharp canines with which the mature baboon can inflict severe wounds do not erupt until the animal is about four years old. All through his earlier years the baboon has been engaging in rough play, fighting with his contemporaries, but he does not have the weapons to make his play dangerous. When he matures sexually and the teeth are fully erupted, his muscles also are ready for the serious fighting that will determine his status in the troop. By that time he also has fully mastered the techniques of fighting. It all works out together.

The young females at first also take part in the romping. However, by the time the male play grows rough, the young females often show more interest in the infants of the troop. Most young females are experienced infant handlers before they have any young of their own.

Play has been stressed as the time in which primates learn to control their behaviors. Youth is the time for intellectual development; the young learn what to eat and what to avoid. They learn social behaviors and the control of emotions. The control of emotions is one of the ways in which humans differ most from their nearest living relatives. We are so used to living in a world in which people's emotions are under control that we think of uncontrolled behavior as criminal. But the sexual or aggressive acts we think of as unacceptable are common in the societies of monkeys and apes. For example, according to Galdikas, *all* old male orangutans have scars from fights with other males.[17] Chimpanzees may kill other chimpanzees and eat slaughtered infants. The Gombe group of chimpanzees killed all the males of a smaller group.[18] An audience at a lecture was horrified when Dian Fossey described how gorillas sometimes killed each other, and, again, sometimes ate dead infants.

Such aggressive behaviors are made possible by anatomy, and the occurrence of the aggressive acts was predicted from the anatomy long before the aggressive events were observed.[19]

Human emotions are controlled by the cortex of the brain (see page 169), language, and customs in a way that is quite new, and whether or not our apelike ancestors of millions of years ago were aggressive would tell us nothing about the problems of today.

Recently, there have been numerous attempts to describe these ancestors as aggressive or cooperative, but these two kinds of behavior are not mutually exclusive opposites. In warfare, people may be both extremely aggressive and highly cooperative. On an African field trip one of us saw a group of over 180 baboons crossing an open field. Two dogs ran toward the group barking. The females and smaller baboons retreated slightly, and some two dozen big male baboons stood between the dogs and the other baboons. The dogs turned and fled. Yet four of these big males had been fighting among each other just a short time before. One cannot speak of aggression or cooperation without regard to time, place, and the reasons for the actions. Among the nonhuman primates, aggressive actions, actions leading to one animal damaging another, are common. Cooperative actions, actions in which one animal helps another, are rare. The reasons for this may be in large part the problem of communication without language.

Numbers and Breeding

The groups of chimpanzees and gorillas are small. There is great variation locally, so that one to four dozen may reflect the situation better than exact numbers. Groups are larger in many kinds of monkeys, but even so inbreeding (mating within the group) would be prevalent unless some animals left the group. Usually, it is the males who leave and join a new group, but in chimpanzees it is the females. Apparently, in gorillas both males and females may shift. Breeding out is a fundamental rule of primate societies.

Inbreeding is not the only problem for a very small social group; worse, the next generation may be all of one sex. The bands of our ancestors must have been small, and pure chance would have made for impossible situations. For example: In a group of forty there might be nine couples. About three infants a year would be produced, and possibly one of the three would survive to maturity. The chance was about one in eight that in a three-year period young adults might all be male or all female. Certainly, departures from an equal sex ratio in such a small group would be common, and, if continued, the existence of the group would be threatened.

It has been calculated that to maintain the 50/50 ratio of males and females needed for a smooth social life at least 100 pairs would be necessary, which would require a population of at least 500. The australopithecine bands on the savannah were quite unlikely to reach such a size. To correct the inevitable imbalance, mates would have to be found in neighboring groups. In a hunting society, the hunters would encounter them. Though the disturbance of game cannot be tolerated, many hunting societies live peaceably side by side. After some intermingling, it would be easy for a mother to suggest to her son that he might find a mate in the group from which she herself had come. Ties of custom and kinship would spring up.

The finding of mates and the production of babies under the particular conditions of human hunting and gathering favor exogamy and also the incest taboo (to prevent the birth of infants before there is a male economically able to support them). The assumptions behind this argument are that social customs are adaptive and that nothing is more crucial for evolutionary success than the orderly reproduction of the number of infants that can be supported. This argument also presumes that at least under extreme conditions these necessities and reasons are obvious to the people involved, as infanticide attests.

The biological necessity for members of very small groups to seek mates outside the group was supplemented by the demographic necessity of breeding out to find mates in small human groups. So exogamy and the supporting incest prohibition may be one of the few human customs clearly rooted in the behaviors of nonhuman primates. It is odd that the very customs that were supposed to be the basis for the difference in social behaviors between ape and human may turn out to be those that are in fact most similar.

Of course human beings elaborate everything in a way nonhumans do not, and the rules of exogamy are no exception. There is a vast difference between kinship systems, which indicate preferred marriages, and outbreeding, which involves mating any member of a different group. Probably the difference between an ape mating system and human kinship depends on language and cognitive development, to which we will turn in the next chapter.

6

THE THINKER

In beginning this last chapter we wanted to stress the continuity of thought about human evolution and add some recent developments. The continuity has been stated by J. S. Weiner in the following paragraph and then in a quotation from Darwin's own writing.[1]

> [Darwin] argued therefore that the early hominid progenitor must have been ape-like in having a small brain and that the brain would have expanded subsequently under the influence of natural selection. Darwin went on to claim that the adoption of the erect posture was in fact the key to understanding not only the enlargement of the brain but the total transformation of the ape form into the human form. The primacy of bipedalism is elegantly argued.

> Man could not have attained his present dominant position in the world without the use of his hands. . . . But the hands and arms could hardly have become perfect enough to have manufactured weapons, or to have hurled stones and spears with a true aim, as long as they were habitually used for locomotion and for supporting the whole weight of the body, or, as before remarked, so long as they were especially fitted for climbing trees. . . . From these causes alone it would have been an advantage to man to become a biped; but for many actions it is indispensable that the arms and the whole upper part of the body should be free; and he must for this end stand firmly on his feet.

The next steps in the argument are these:

> If it be an advantage to man to stand firmly on his feet and to have his hands and arms free, of which, from his pre-eminent success in the battle of life, there can be no doubt, then I can see no reason why it should not have been advantageous to the progenitors of man to become more and more erect and bipedal. They would thus have been better able to defend themselves with stones and clubs, to attack their prey, or otherwise to obtain food. The best built individuals would in the long run have succeeded best and have survived in larger numbers.
>
> As the progenitors of man became more and more erect, with their hands and arms more and more modified for prehension and other purposes, with their feet and legs at the same time transformed for firm support and progression, endless other changes of structure would have become necessary. The pelvis would have to be broadened, the spine peculiarly curved and the head fixed in an altered position, all of which changes have been attained by man.
>
> The free use of the arms and hands, partly the cause and partly the result of man's erect position, appears to have led in an indirect manner to other modifications of structure.
>
> The early male forefathers of man were probably furnished with great canine teeth but as they grad-

> ually acquired the habit of using stones, clubs or other weapons for fighting with their enemies or rivals, they would use their jaws and teeth less. In this case, the jaws, together with the teeth, would become reduced in size.
>
> As the jaws and teeth in man's progenitors gradually became reduced in size, the adult skull would have to resemble more and more that of existing man.
>
> As the various mental faculties gradually developed themselves the brain would certainly become larger.[2]

The record of the fossils fits what Darwin had predicted. There was, however, a long period in which Darwin's influence waned, and alternative theories were widely believed. One of the most influential of these theories, especially supported by the anatomist Elliot Smith and Arthur Keith, was that the evolution of the brain had separated ape and human. This theory dominated evolutionary thinking from about 1900 until the late 1940s when the pelvis and limb bones of *Australopithecus* were discovered by Robert Broom.

The idea that the large brain was early and primary in human evolution laid the background for one of the most successful forgeries in the history of science. In 1912 Charles Dawson and Arthur Smith Woodward announced the discovery of Piltdown, a remarkable human skull. The braincase was large and modern, but the jaw resembled that of an ape. There was immediate controversy as to whether the two could have belonged to the same individual. Elliot Smith pointed out that this was what should have been expected—that it proved the brain had evolved long before the human face attained its modern proportions. The debate appeared to be settled by Dawson's discovery of a second Piltdown "fossil," a fragmentary find that included a piece of skull and a tooth, supposedly confirming the combination of features seen in Piltdown 1.

Some years before, in 1890, the Dutch physician Eugène Dubois had unearthed a fossil of a very different kind of early creature from a riverbank in Java. A jaw fragment and a molar tooth had looked near human, but the cranium was low and flat, not at all as high as that of a human. The suggestion that a human ancestor could have had so low a skull and so receding a forehead outraged both lay and scientific worlds. The dissenting uproar was so great that Dubois withdrew some of his Java finds from scientific exhibition and locked them in a strongbox for nearly thirty years.

Under the circumstances, the Java finds did not generally upset the theory of how humans evolved. The Piltdown forger did not redesign the forerunner he was creating; he still gave us the kind of ancestry many expected.[3] Although many scientists worried about an apelike jaw with so high a head, it was difficult to quarrel with what appeared to be the actual record from the English ground.[4]

Even in 1925, when Dart announced the discovery of the Taung skull, the old theory still prevailed. Part of the refusal to accept the six-year-old australopithecine stemmed from the continuing belief that this could not be the way it had happened. A human brain might go with a subhuman body, but surely it could not be the other way around. The first verdict was that the South African creature was an ape.

Only as the discoveries began to come in from Sterkfontein and Kromdraai and reports belatedly issued from the laboratories did fact overcome belief. Then the unwelcome truth emerged—humans had begun the long upward climb with a brain no larger than the ape's. Later studies showed that the human brain remained apelike in size long after the human line separated from the apes.

Le Gros Clark, who made careful studies of the australopithecines, readily conceded that the small braincase combined with massive and projecting jaws gave a superficial resemblance to the skull of anthropoid apes. With the benefit of hindsight, the English scientist wondered a little that such apishness should have been so hard for many to accept and commented:

> It was to be expected that in still earlier and more primitive representatives of the hominid line of evolution the brain would be even smaller, perhaps hardly exceeding simian dimensions, and that the jaws would similarly be more massive and projecting in a simian fashion.
>
> In retrospect it may seem surprising that students of fossil man were not ready to accept the obvious inference that the earlier prepithecanthropine stage of hominid evolution must have been characterized by small braincases and massive jaws, for it was generally held that hominids and pongids were derived from a common ancestral stock and such characters of common inheritance would certainly be shown in the initial stages of hominid evolution and perhaps persisted for some time.[5]

The first observations were confirmed. The australopithecines had not advanced in brain size beyond their ape ancestors, and the human start had been made with a small brain.

The relationship between locomotion and brain size has been further clarified by the discoveries at Laetoli and Hadar. The fossil footsteps, dated at 3.6 million years, show fully human locomotion. Surely, bipedalism must have been evolving for some time before that, so bipedalism is at least 2 million years older than the first of the larger brains that appear in the record some 2 million years ago. It is interesting to note that when Le Gros Clark first went to Africa to examine the original specimens of *Australopithecus* he thought they were apes rather than forms in a stage in human evolution.[6]

The evidence had piled up impressively. Le Gros Clark's close study of *Australopithecus,* and his meticulous comparison of it with the great apes and *Homo,* convinced him that the australopithecines were hominids rather than pongids.

Patterns: Relationship Between Form and Function

Le Gros Clark not only described the apes and australopithecines; he also developed new ways of making such comparisons. He showed that comparing anatomical features, or lists of such items, might be very misleading, and he devised the concept of "total morphological patterns" to clarify the comparisons. For example, the size of the brain is directly related to the size of the braincase; the size of the brain and the surrounding braincase may not be treated as if they provided independent kinds of information. Likewise, the size and form of the face is clearly related to the size of the teeth.

Le Gros Clark did not describe what he meant. We need to examine his fundamental concept of total patterns before considering the evolution of skull or brain. Figure 6.1 shows the skull of a female gorilla and a human being sectioned in the midline. The great size of the human brain is clearly evident. The pressure of the fluid surrounding the brain (cerebrospinal fluid) induces the bones covering the brain to grow, and this is why the capacity of the skull accurately reflects the size of the brain. Unfortunately, since the brain is separated from the bone both by the fluid and by protective coverings, the detailed anatomy of the brain is not reflected in the bone.

The patterns in the face are much more complex, but the general method of comparison is easily illustrated. Figure 6.2 shows the developing teeth in the skull of a child. The face is full of teeth. The teeth and supporting bone are in a functional relation, as are the brain and its bony case. The erupting teeth stimulate the formation of the bone that supports the teeth, and experiments show that if the teeth are removed long before they erupt, the bones do not form. As is well known, in old age, after the teeth have gone, the formerly supporting bone disappears. The general situation is clear and simple, but it is complicated by the

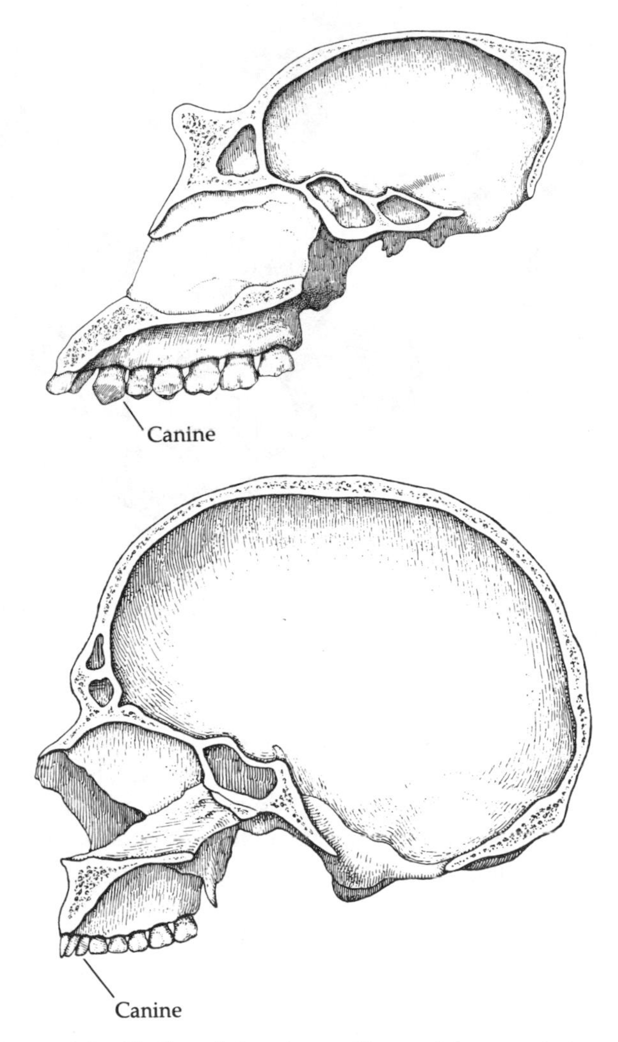

Figure 6.1 Skulls of female gorilla and human being sectioned in the midline to show the relative proportions of brain and face.

different kinds of teeth, the muscles associated with them, and varying functions.

For example, Figure 6.3 shows the palate and upper teeth of a robust australopithecine. The molar teeth

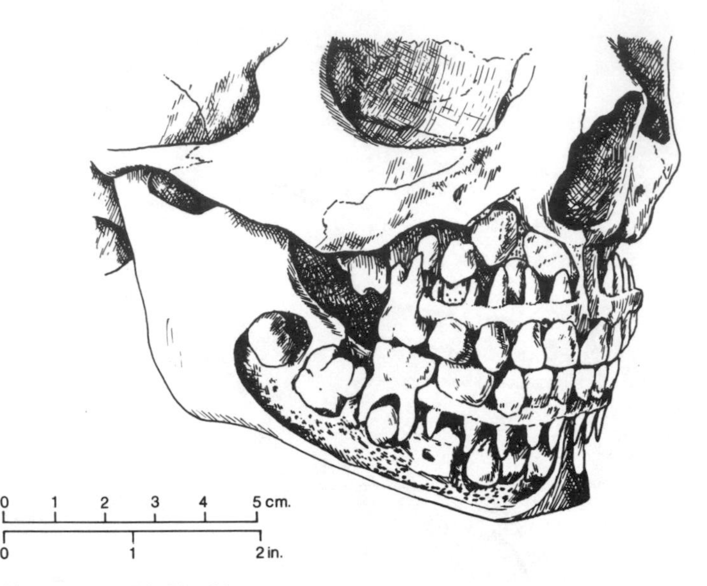

Figure 6.2 Skull of human child. Note that the face is full of developing teeth.

are very large; but the front teeth, incisors, and canines are very small. The gorilla shows the opposite—the front teeth, particularly the canines, are very large. The muscles that move the jaws begin all the way back at the neck, where there is a big crest in the gorilla. In the robust australopithecine there are crests that show that the muscles were further forward.

To interpret the facial patterns, the whole structure must be related to the function of the teeth. The very small anterior teeth of the robust australopithecine suggests that these were not very important; dentition was adapted primarily to crushing large food objects. In the gorilla the anterior teeth are used in feeding and fighting. The canines are particularly important, and if a male animal fights with its face, it must have powerful neck muscles. The major sex differences in the canine teeth and skull of male and female gorillas supports this interpretation.[7] But how did the robust forms

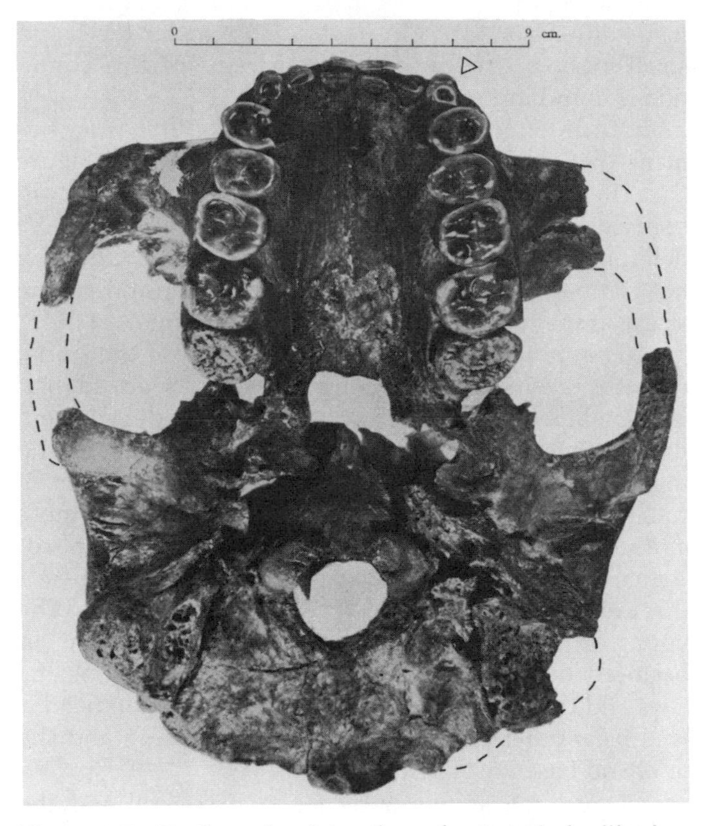

Figure 6.3 Teeth and palate of a robust australopithecine. Note that the canine is far smaller than the adjacent tooth.

fight? The best guess is that they were using tools and that their ancestors had been using tools for a long enough period of time for the genetically determined reduction of the canines.

The essential point is that it is complexes of structure that help us understand the relationship between function and evolution. If the robust australopithecines had been constructed like apes, we would expect them to have bigger front teeth, longer faces, bigger neck muscles, and a longer cranial base. From this point of view, the robusts were ultrahuman, not primitive. Humans differ from apes in having large molar teeth and small incisors and canines; the extreme of this condition is found in the robusts.

It has often been stated that apes differ from humans in the shape of the palate, as shown in Figure 6.4. This is correct, but looking at the figure one can see why. The big incisors and huge canines of the ape determine the form of the front of the palate. One may not compare palate form, facial projection, and canine size as if they were separate, unrelated things.

The human head is balanced on top of the spine, and this position is frequently treated as an adaptation to bipedalism. Posture has been deduced from the characteristics of the skull. But it takes almost no muscle to balance the head. Resisting the strains of roughhousing, boxing, or wrestling takes vastly more strength than the amount required to balance the head. Actually, some fully quadrupedal monkeys have heads that are as well balanced as those of human beings. The young of many apes and monkeys also have well-balanced heads; the head becomes unbalanced as the jaws and the base of the skull grow. The human head is well balanced because the brain is very large and the jaws and base are short.

Figure 6.5 shows a section of the skull and the neck of a gorilla. The spines of the cervical (neck) vertebrae are long, and this feature has been called an

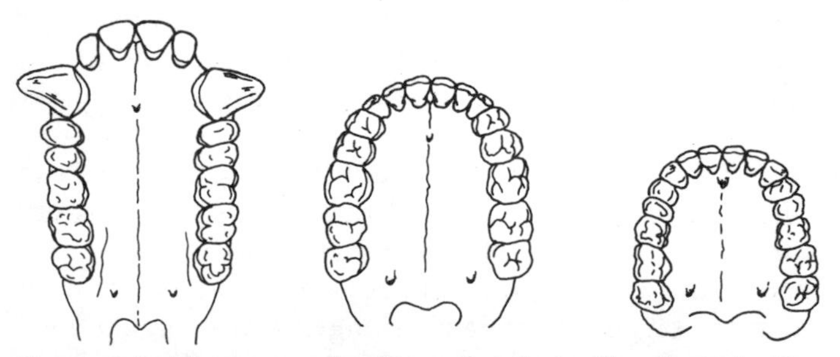

Figure 6.4 The upper dentition of male gorilla, *Australopithecus,* and *Homo sapiens* (from left to right).

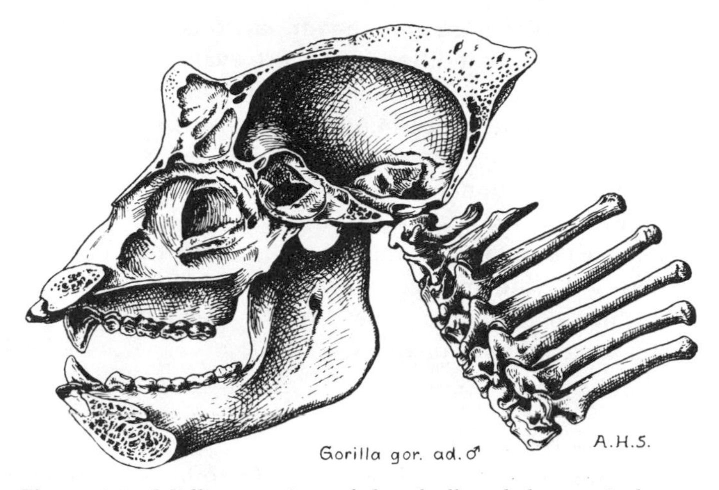

Figure 6.5 Midline section of the skull and the cervical vertebrae of a wild adult male gorilla.

adaptation to balancing the head. The principal muscles (semispinalis capitis) that balance the head do not, however, arise from those spines; the major arm muscles arise from the spines.

The point is that the bones were parts of living beings. To assess adaptation and evolution we have to do our best to see how the parts worked during the lifetime of the individuals. This is where Le Gros Clark's concept of *pattern* is essential. The degree of change from ape to human cannot be estimated without it.

Figure 6.4 showed a comparison of the palate of male gorilla, *Australopithecus*, and *Homo sapiens*. The differences would seem far less if a pygmy chimpanzee were used as the basis for comparison.[8] Differences would be further reduced if the earliest australopithecines were used. Figure 4.7, page 110, shows the lower jaw of one of the specimens found at Laetoli. The canine is larger than in later forms, and the first lower premolar shows primitive features. So both the way the differences were described and the choice of which specimens were used in the comparisons affected the way the ape-human differences were evaluated.

Comparison with the adult male gorilla made the differences between apes and humans seem very great and a short time of separation nearly impossible. Comparison by functional patterns of the pygmy chimpanzee and the earliest australopithecine fossils makes the differences seem much smaller, of an order that is compatible with the molecular evidence and a relatively short time of separation. Remember, "relatively short" is still some millions of years!

The Brain

If the important early events in uniquely human evolution were bipedalism and the use of tools, the significant late events were language and intelligence. We are so immediately and deeply involved in these abilities that it is very difficult to picture a world without them. Imagine not being able to communicate about what happened yesterday or to make a suggestion about tomorrow. Imagine not being able to count, measure, or transmit a complex tradition. Religion, medicine, social customs, and technical progress all depend on language and cognitive abilities, and the evidence is that these took their modern form only at the very end of human evolution.

If we consider the fossil record in the most general terms, the small-brained bipeds existed from before 4 to about 2 million years ago. They appear not to have been more numerous or more successful than many other forms of life. Then in a period of some 500,000 or 1 million years, Oldowan stone tools developed, and the brain doubled in size. Probably these are two aspects of the same events that were in a feedback relation to each other. Beginning about 1.5 million years ago *Homo erectus* made stone tools of the Acheulian tradition. For hundreds of thousands of years humans made the same kind of tools, and although some tools were skillfully made there is no evidence of great progress. By 100,000 years ago progress accelerated a little. Neanderthals buried their dead. Tools became more diversified and complex. It was only in the last 40 thousand years—only in the last 1 percent of human evolution—that the rates of change and kinds of behavior we think of as normal appeared in the archaeological record. (See Table 6.1.)

If we compare this time record with the structure of the human brain, we see that there is a remarkable similarity. Figures 6.6 and 6.7 show that the area for hand skills is very large, far larger than in any ape. As Figure 6.6 shows, the part of the cortex that controls the thumb is as big as that for the rest of the hand. In the apes the thumb is small and weak. They do not have the large, powerful muscle that flexes our thumb and makes possible a powerful grip.

Putting this in historical perspective, the evolutionary success of object using led to selection for larger, more powerful thumbs and a much larger representation of the thumb in the cortex. The cortical representation makes hand skills possible and pleasurable to learn. Think of all the activities that involve fingers compared with those that involve toes and compare the

Table 6.1 *Size of the brain measured by the capacity of the braincase. Measurements are in cubic centimeters. The numbers are* representative *capacities, that is, the most commonly occurring. The extremes are not included.*

Homo sapiens 1,100–1,500	No evidence of increase in the last 100,000 years
Homo erectus 800–1,000	Skulls from Africa, Java, China, Europe
Transitional forms 650–750	*Homo habilis*, if a name is wanted
Australopithecus 400–500	Capacity of large form slightly higher
Chimpanzee 350–450	The same for the orangutan
Gorilla 450–550	But gorillas are much larger than the other apes

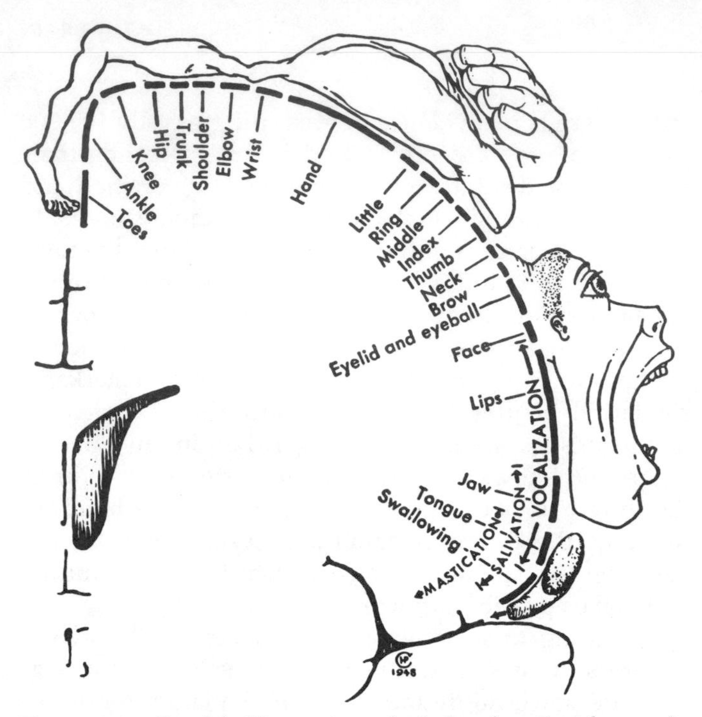

Figure 6.6 Graphic illustration of relative functional areas in human motor cortex.

amount of cortex controlling the two kinds of activities. The very structure of the brain mirrors what has been important in human evolutionary history.

There are two important qualifications to what has just been said. First, the distorted human figure shown in Figure 6.6 makes a point correctly but in greatly oversimplified form. The areas of the cortex are not sharply distinguished but are emphasized in different ways. For example, about 25 percent of the responses in the motor area are sensory. Areas for hand and thumb overlap, and mapping makes the areas appear more distinct than they really are. Second, the whole brain evolved, and changes in the cortical areas are correlated with those in the underlying structures. For example, the cerebellum is essential to the integration of motor activity. As the selection for cortical control advanced, changes in the cerebellum and other

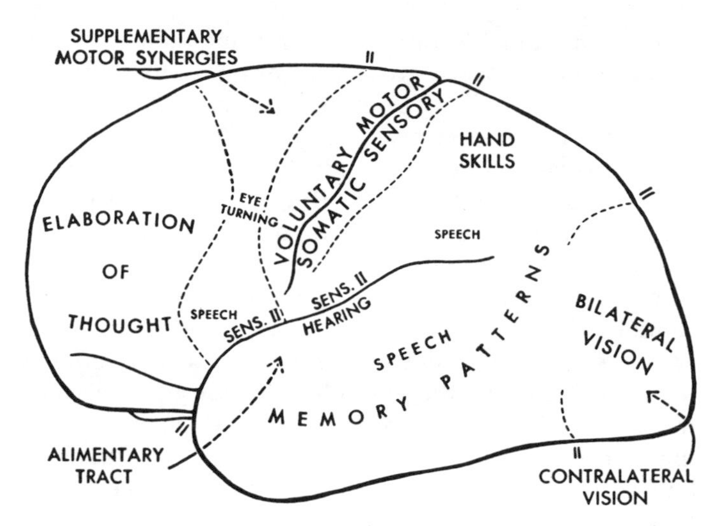

Figure 6.7 Functional areas of the human cortex. Note how much is concerned with speech.

parts of the brain were also favored. The whole functioning brain is extraordinarily complex and, even today, is only partly understood.[9]

Judging from hand skill as reflected in the archaeological record, the brain doubled in size and human skill increased enormously from 2 to 1 million years ago. This was the period of major change as far as gross size is concerned. But the comparison of the brains of contemporary apes and humans shows that there has been internal reorganization, too, which is not revealed in the fossil record. Saying that the brain has doubled may be a gross underestimation of the amount that the brain has evolved.

Communication

Communication is essential to any kind of social system. Even the apparently solitary male orangutan starts the day with a booming cry that tells other orangutans where he is. Sounds of this kind are common among the primates and many other mammals. Where primates live in social groups, communication is much more complicated. The animals must judge others' emotions, which are conveyed by gestures and sounds. Bluffing is very important, and all the apes have biological structures adapted for bluffing. Gorillas, for example, pound their chests. Chimpanzees charge, hoot, and throw objects. In orangutans there are very large sacs connected with the larynx, and these make the territorial noises possible. When an ape and some monkeys are demonstrating aggressively, the hair on the heads and around the shoulders stands up. This has the effect of making the creature look two or three times its normal size, and being subject to a sudden bluff of this kind can be a scary experience. (See Figure 6.8.)

Primates constantly communicate moods and information about food, dangers, or social states. After

Figure 6.8 Baboon threatening. The hair stands on end, magnifying apparent size of animal.

observing baboons in the Amboseli Reserve in East Africa, Stuart Altmann wrote, "The coordination in a large troop is a wonder to behold."

But in spite of the richness of the combinations of gestures and sounds, all the nonhuman systems of communication are extremely limited. Other animals cannot communicate about the past or the future. Most communication (except for warning cries and territorial noises) is primarily gestural, and the animals must be able to see each other in order for communication to take place. An even greater limitation is that communication is confined almost entirely to emotions. All the primates can communicate fear, but only humans can say what they are afraid of. Some monkeys have sounds indicating danger from above (action, drop to lower branch) or from below (action, climb up), but they cannot indicate the nature of the danger.

Figure 6.9 pictures a gorilla's facial muscles, and we can see that they look very much like our own. Gorillas can make many sounds, but they seem to be able to communicate very little. Experiments with monkeys show that both sounds and facial expressions are controlled by primitive parts of the brain, parts that also control emotions, and that the cortex minimally controls sound or expressions.[10]

There have been many attempts to teach monkeys or apes to speak. All failed. In their studies of chimpanzee communication R. Allen and Beatrice Gardner wisely shifted their efforts from sounds to gestures. Their success in this endeavor opened a whole new area of research, and many other scientists have followed their lead. It is now clear that all the apes can learn to communicate by gestures and that several hundred gestures may be learned. This fact has opened the way to a far greater appreciation of the cognitive abilities of the nonhuman primates but has also demonstrated the fundamental difference between ape and human abilities.

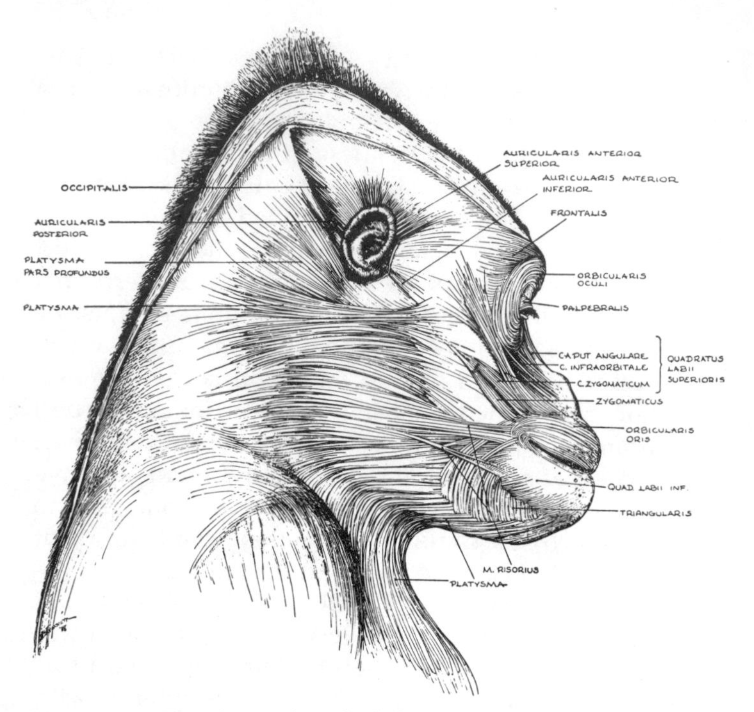

Figure 6.9 Facial muscles of adult male gorilla.

Speech

Human speech is something new. It is controlled by the cortex. The sounds of monkeys and apes are controlled by primitive parts of the brain, parts that primates share in common with many other mammals. The vocalizations of nonhuman animals are fundamentally different from human speech.[11]

Human beings learn to speak so easily that it is almost impossible to prevent learning of this kind. If a child is deaf, it will make great efforts to communicate by gestures and will easily learn many more gestures than any ape. Thus, although the old channel for communication remains, the new speech channel has been added. The new structure that makes learning speech so easy is in the cortex. In hundreds of operations performed on human beings under local anesthesia, the brain has been explored as an essential part of planning operations to relieve a wide variety of complaints (e.g., tumors, trauma, epilepsy). Each letter in Figure 6.10 represents a point where a surgeon found that an exploring electrode interfered with speech. On the dominant side, usually the left, the speech area is very large; on the opposite side, usually the right, only probing in the motor area causes speech disturbances. Comparison of left and right sides shows that the differences between ape and human is both in the quantity of cortex necessary for speech in human beings and in the lateralization of this ability to one side. (See Figure 6.11.)

In a general way, these aspects of the human brain had been known for a long time, ever since Paul Broca, a French surgeon, discovered a speech center in the 1860s. Recently, Roger Sperry of the California Institute of Technology extended our understanding of the lateralization in brain functions far beyond speech.[12] In some patients, it has been necessary to cut the corpus callosum, a massive bundle of 250 million fibers, which connect the two lateral hemispheres of the brain. After such an operation, the patient has two remarkably independent and different brains. Information may be fed into either half. The left half can then speak and is self-conscious; the right half can think and solve problems (especially of forms or music) but cannot speak or self-consciously consider the problems it is solving. Normally, the various senses send information to both sides of the brain, and the activities of the cortex are coordinated through the corpus callosum, the interhemispheric connection previously mentioned. In the individual with divided hemispheres, however, the functional emphasis of each side of the brain may be studied separately.

In rare cases, one whole cerebral hemisphere must be removed. If this happens before five years of age, the language function locates on the remaining

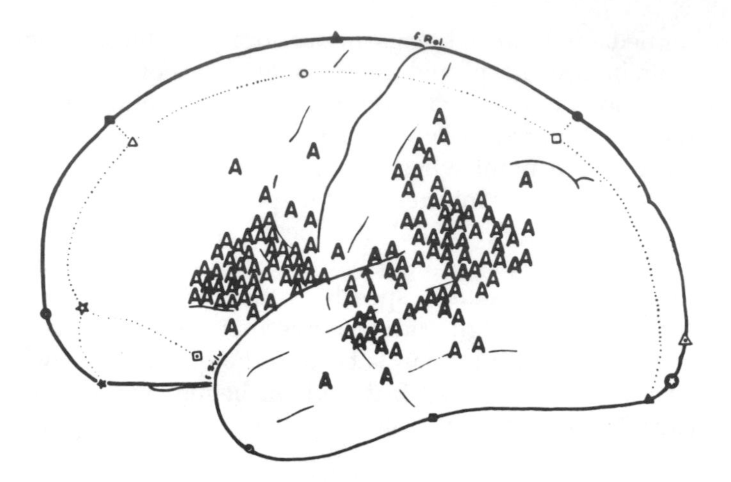

The Aphasic Types of Responses

Figure 6.10 Left side of human brain. Each letter marks a place where speech was arrested.

side. Further, the single hemisphere performs cognitive functions remarkably well. Michael S. Gazzaniga,[13] who has been a leader in studies of the divided brain, points out that the single hemisphere may perform IQ tests as well as the intact whole brain. This ability clearly shows that it is the neural structure that makes the human brain unique. The single-hemisphere person has only one half of the mass of brain of a normal person, yet such a person's cognitive functions are nearly normal.

Given these facts about the human brain and how it functions, with regard to evolution and the fossil record, one can easily see why the study of the size of the brain alone is so inconclusive. If the most important changes have been in the neural structure, and these changes do not appear in the fossil record, the size of the fossil skulls presents no direct evidence of which one could have spoken and which could not. This is the case even though speech requires a large amount of

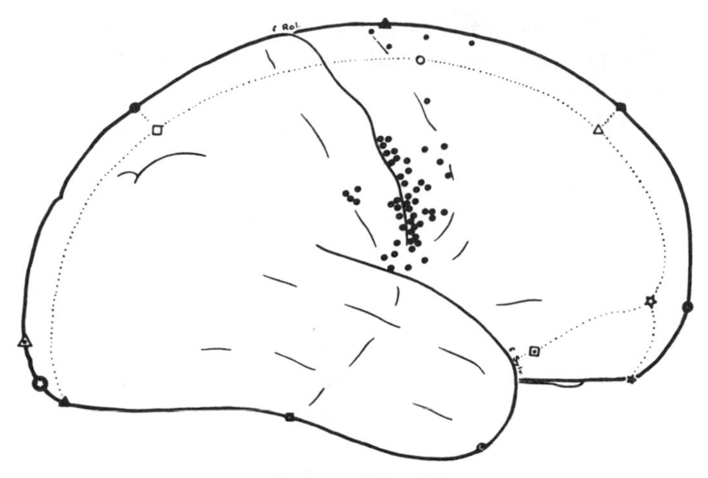

Arrest of Speech

Figure 6.11 Right side of human brain. Arrest of speech occurs only at these few points.

brain. Nor does lateralization show in the fossils; the degree of difference between the right and left hemispheres came as a surprise even to those who had studied the brains of contemporary humans in great detail.

Granting these difficulties, we may learn more about evolution from the structure of contemporary human brains than from the fossil record. For example, human hand skills depend on the greatly enlarged neural structure of the cortex, the cerebellum, and some other parts of the brain. In the fossil record the brain doubles in size during the period in which simple Oldowan stone tools evolved into the Acheulian tools that are very difficult to make. Combining the study of the brain with the fossil record suggests a functional association that might account for part of the increase in brain size. From 1 million to 100,000 years ago stone tools changed very little, which may indicate that the toolmakers were not as intelligent as the later human beings. By the time of the Neanderthals, the brain was a little larger than it is today. Was the decrease due to lateralization and language?

With *Homo sapiens* the rate of evolution accelerates. We think that this acceleration, the rate of change, was the result of language. Of course there must have been more primitive forms of acoustic communication. As new methods of studying the brain are developed, and as archaeology improves so that the intellectual tasks our ancestors faced can be more precisely stated, it is possible that the evolutionary forces that built the modern brain can be stated more precisely.

To summarize the present situation: human beings easily learn to walk, to speak, to manipulate, to control emotions, to be social, to be intelligent. And it should be emphasized that humans elaborate everything in a way that is found in no other animal. They do not just walk and run, but do inventive dances, do complex gymnastics, race to win, and play intricate games. They do not just make sounds; they make songs, poetry, speeches, prayers, and a multitude of symbolic expressions. Food and sex are complicated in ways completely beyond what might be expected from the behavior of any other animal. Manipulation is not used only for technical matters but also for art in all its many forms. Humans easily learn and enjoy this enormous extension of biological behaviors into the myriad forms of culture.

Judging both from the archaeological record and from the study of the last few hunter-gatherers, language and complex human ways of life were present before the agricultural revolution that began some 10,000 years ago. In the last few thousand years history accelerated, and human beings learned to make boats, cross large bodies of water, conquer the Arctic, and enter the New World. All contemporary peoples have the biology that makes language learning easy, and it seems apparent that this ability (speech, plus related cognitive factors) is the basis of the most distinctive behaviors of *Homo sapiens*.

But having the ability to speak does not mean that we understand what speech is, which should not be surprising. We learn to walk without knowing anything about the muscles, joints, and neural control that make this behavior possible. Perhaps only a few surgeons and specialists in physical education need to know about the biology of locomotion, but we have paid, and continue to pay, a very heavy price for not understanding the nature of speech.

The situation may be clarified by comparing human speech with the communication of the nonhuman primates in terms of *function*. Human speech may communicate an almost infinite number of meanings. It is the adaptation that more than anything else differentiates human behavior from that of the other primates. If new words are needed, they can be coined. All human languages are open. We can talk about the world in which we live. In nonhuman primate communication, messages are limited, fixed in number, and are primarily concerned with the inner state of the animal sending the message. Primates cannot talk about the world but only state, in effect: "I am afraid, desirous, worried, comfortable," and the like. Even these messages cannot be qualified by such simple additions as, "You are afraid; I was afraid yesterday." Nonhuman communication has no nouns and no syntax.

The new human adaptation is made possible by a sound code. Short sounds (phonemes) that have no meaning may be combined into meaningful units (morphemes, words). The sound code has the characteristic of codes in general in that a few elements may be combined in an almost infinite number of ways. In nonhumans there is no code. There is no beginning of speech. The phonetic code is both appreciated and controlled by the cortex on the dominant side of the brain, as shown in Figure 6.10. In contrast, the sounds of the nonhuman primates are controlled by primitive parts of the brain.[14]

Clearly, the human cortex-sound, code-word system is new and limited to human beings. The primitive-emotional system is old and widespread among the vertebrates; in this system, gesture is more important than sounds. Apes are highly intelligent animals, and the old gestural communication system may be greatly improved by human training, as we have seen. This not only gives us a much greater understanding of the ape's behaviors but also shows the basic cognitive similarity of ape and human.[15]

Semantics

The brain recognizes the patterns of sounds, and the meanings of these patterns are arbitrary. The child learns to associate certain patterns in its language with particular meanings, but in another language the same sounds might have totally different meanings. Even the same sounds in the same language may have very different meanings, and this may cause confusion. For example, the word "language" may be used to mean speech and related cognitive factors. In this usage, language is limited to human beings and is made possible by unique features of the human brain. Or "language" may be used to describe communication in bees, dolphins, or chimpanzees. Both usages are common, and neither is incorrect, but it may be very misleading to slip from one to the other, and even more so to say that the gap between ape and human has been narrowed by saying that apes have language. Apes do not speak and humans do. To keep this issue clear, we have used the words "speech" and "talk" and have stressed the structures of the brain that make these behaviors possible.

Importance of Speech—Costs of Ignorance

Human speech fundamentally changes the nature of human experience. Human experience is different from that of any other animal because, prior to the evolution of speech, the behavior of our ancestors was determined by their biology and by observational learning. That is, success or failure was based on personal experience. By watching the behavior of monkeys and apes, we have learned how limited such information actually is. There is no way they can discuss the events of the past or consider the possibilities of the future; there is no way for them to consider alternatives or the improvement of patterns of behavior. Limited to personal experience and the observation of a very few other animals behaviors must be largely preprogrammed, or genetically determined. There is no way to incorporate the experiences of many generations and thousands of individuals into useful, learned traditions. It is speech that makes possible the human way of life.

As noted previously, human beings do not know that they are recognizing sound patterns that may be described as composed of series of meaningless sounds, phonemes. It appears to human beings that words are the units, and this both underemphasizes sentences and makes it impossible to analyze words. When people first began to write, symbols stood for words; in this system something on the order of 10,000 symbols are required for even moderately useful communication. This was the case for Chinese; a very large part of a scholar's time was devoted to learning the thousands of symbols. After the first writing more than 2,000 years passed before human beings discovered that a few symbols for the basic sounds could replace all the symbols of words. The invention of the alphabet in the Near East was one of the great events in intellectual history; if people had recognized from the beginning that in writing they were using a sound code, millions of hours of human effort could have been saved.

Furthermore, understanding the way phonemes, the basic sounds, are related to the sound patterns (words and sentences) should lead to a clear understanding of spelling. Our schools waste millions of hours of students' time by teaching archaic spellings rather than clarifying the fundamental nature of language. Given the nature of modern human biology, all human beings can enjoy learning language, and all normal people will make great efforts to do so. Learning a second language is no problem, provided that the learning occurs in a natural situation. It must be clear to the child that the second language is useful, and it must be practiced by talking with native speakers. Our schools provide neither the motivation nor the opportunity to learn a second language effectively. The partial learning of languages that are useless to most students has caused, and continues to cause, an incredible waste of human time and effort.

Language and Facial Expression

As noted earlier, the sounds of the nonhuman primates are limited in number and primarily convey the emotional state of the animal making the noises. The removal of large amounts of the cortex of the brain does not affect the production of sounds. In human beings comparable lesions result in facial paralysis. The same is true of facial expressions; they mirror emotional states and are not altered by insults to the cortex. In human beings damage to the cortex, even in quite small areas, may cause facial paralysis. Just as in the case of speech, the fundamental biological difference in neural control is reflected in the ease of learning. With control by the cortex, human beings can easily learn facial expressions as well as which particular ones are socially appropriate in different situations. The primitive relation of emotion to facial expression may be retained, but it is greatly modified by learning. For example, in a child crying may mean sorrow, but older people may learn to cry when it is expected, even if they do not feel the emotion of sorrow at all.

From an evolutionary point of view, we might

speculate that speech depends on short sounds that require precise control of the movements of pharynx, tongue, and lips, and on expanding cortical control of the sound production mechanisms. An evolutionary consequence of this was a much greater cortical control of the muscles of facial expression. As can be seen in Figure 6.9, page 172, which shows the facial muscles of the gorilla, the muscles that control facial expression are essentially human. The differences lie in the nature of the neural control of the muscles.

Learning and the Brain

A number of years ago David Hamburg wrote:

> The adaptive function of primate groups should alert us to look for processes in the individual that facilitate the development of interindividual bonds. In seeking such processes, we may find useful guidance in the principle that *individuals seek and find gratifying those situations that have been highly advantageous in survival of the species*. That is, tasks that must be done (for species survival) tend to be quite pleasurable; they are easy to learn and hard to extinguish. Their blockage or deprivation leads to tension, anger, substitutive activity, and (if prolonged) depression. . . .
>
> Society is not composed of neutral actors but of emotional beings—whether we speak of baboons, chimpanzees, or man, emotion lies at the core of the social process. We fear for ourselves, a few loved ones, and the infants of the species. We are positively bound deeply by a few relations. Threat to these relations is equivalent to an attack on life itself. From the standpoint of the species, these are the critical relations for survival. The physiology of emotion insures the fundamental acts of survival: the desire for sex, the extraordinary interest in the infant, the day-to-day reinforcement of interindividual bonds.
>
> From the standpoint of the individual, the greatest satisfaction and fulfillment come from relations with the biologically essential few. Even if we colonize a planet, the development of an infant there will depend on the

> presence of another person—a mother, to hold, to love, to give security, to train for the problems ahead. Conquest of the outer world does not free the species from the inner world which made its evolution possible. Social life is rooted in emotion and is basic to survival. A comprehensive human biology must surely take account of one of man's fundamental properties—his social nature.[16]

Hamburg used his psychiatric clinical experiences and the studies of monkeys and apes to reach these conclusions. He saw evolution as the master process that has produced the diversity of life and the emotional, social biped we call *Homo sapiens*. Contrast the attitudes reflected in the quotation to the decision of the United States Supreme Court, which recently ruled that the teacher's right to physically punish students is basic, "even if the spanking is severe, excessive, and medically damaging." The decision of the Court is based on a view of human nature that to make children work, to educate them, physical punishment is justified even if it is extreme. The traditional school tried to educate by discipline, isolation of individuals, and rote learning. But primates learn by social play and clearly seen objectives. The human brain takes years to mature, and human beings enjoy learning and take pleasure in the countless repetition that is necessary for the mastery of social skills.[17]

At present the cry is "back to basics," but the basics are only the educational traditions of the nineteenth century. Going back much further, the study of our evolution suggests a view of human nature very different from that of traditional European culture. Human beings are naturally social, emotional, playful, exploratory, intelligent. This kind of a species is the product of evolution, of past reproductive success. But it was success in a very different world from the one in which we are now living. In a profound sense, we are the primitives, the relics of the days that are dead. But we are also the most successful species that ever lived. In all the billions of years that life has been on this earth and in all the millions of years that the mammals have been dominant, no other creatures learned to speak, to talk about their world, to invent agriculture, and finally to develop modern science.

And here lies our problem. We are basically adapted to a small, flat earth; to short time periods; to small groups; and to the belief that our ways are right. This was the primitive world with all its limitations and difficulties. Now modern science is creating a totally new world of almost limitless possibility. How are these biologically ancient actors to adjust to the new world? Fortunately, human beings are highly adaptable, and this fact holds high promise.

However, the great ability to learn, to adapt, also has its dangers. Our species may easily learn to be social, cooperate, educate, and strive for peace and harmony. Equally, as history shows, the species may learn punitive, aggressive behaviors, treat children in the manner sanctioned by the Supreme Court, and strive for domination. In our view, the past offers no solutions. It gives us a way of looking, a perspective. And from that perspective we can discern three classes of problems, of areas for investigation and ultimately understanding. These are: (1) The individual, the actors in the social systems. How should the individual be treated, educated, and live? (2) The social system. Human beings exist only in social systems, and all the systems that exist today are products of the primitive world. All existing social systems doom many of their actors to frustration, poverty, disease, and war. (3) Scientific progress. For the first time in the history of our earth, perhaps in the whole history of the galaxy, we know how to make progress toward the solutions of the problems.

The lesson of evolution is clear. The study of evolution should free us from the past, from unplanned change waiting for some beneficial mutation.

The billions of people in the world today are here because of the social changes of the last few thousand, and the last few hundred years. The problems of population and energy and the possibility of devastating warfare are the results of the success of recent history. Science has provided food, cured diseases, and helped people in unprecedented numbers to live to old age. History shows how new these problems are, and evolution shows how ancient are the human beings who must solve the problems. The forces that created human beings over hundreds of thousands of years cannot operate effectively in spans of hundreds of years. The study of evolution should make us vividly aware of the contrast between biological evolution, which created our species, and technical progress, which now dominates the world. There is no use in turning to the old evolutionary answers, to processes requiring almost endless spans of time, when the problems requiring solution are here and now. Evolution should help us to see that the world of modern technical culture is new and that new customs are needed for the problems it has created.

Notes

Chapter 1

1. Charles Darwin, *The Origin of Species* (1859) and *The Descent of Man* (1871); reprinted in one volume in the Modern Library (New York: Random House).

2. Adam Sedgwick, Darwin's former professor of geology, in a letter to Darwin, in Francis Darwin, *The Life and Letters of Charles Darwin* (New York: D. Appleton & Company, 1887).

3. Thomas H. Huxley, *Man's Place in Nature* (1863); reprinted as an Ann Arbor Paperback (Ann Arbor: University of Michigan Press), p. 181.

4. Mary D. Leakey, "Footprints in the Ashes of Time," *National Geographic* 155 (April 1979).

5. See Michael H. Day, *Guide to Fossil Man: A Handbook of Human Palaeontology,* 3d ed. (Chicago: University of Chicago Press, 1977).

6. All of these molecular methods are described in Theodosius Dobzhansky et al., *Evolution* (San Francisco: W. H. Freeman, 1977), pp. 276–303.

7. Morris Goodman and Richard E. Tashian, eds., *Molecular Anthropology* (New York: Plenum Press, 1975).

8. Raoul Benveniste and George J. Todaro, "Evolution of Type C Viral Genes: Evidence for an Asian Origin of Man," *Nature* 261 (1976).

9. Morris Goodman, "Protein Sequence and Immunological Specificity," in W. Patrick Luckett and Frederick S. Szalay, eds., *Phylogeny of the Primates* (New York: Plenum Press, 1975), pp. 219–249.

10. Goodman and Tashian.

11. Vincent M. Sarich and John A. Cronin, "Molecular Systematics of the Primates," in Morris Goodman and Richard E. Tashian, eds., *Molecular Anthropology* (New York: Plenum Press, 1975), pp. 139–168.

12. Elizabeth J. Bruce and Francisco J. Ayala, "Humans and Apes Are Genetically Very Similar," *Nature* 276 (1978), pp. 264–265.

13. Marie-Claire King and Allan C. Wilson, "Evolution at Two Levels in Humans and Chimpanzees," *Science* 188 (1975), pp. 107–116.

14. Frank Press and Raymond Siever, *Earth,* 2d ed. (San Francisco: W. H. Freeman, 1978).

15. Robert S. Dietz and John C. Holden, "The Breakup of Pangaea," in *Continents Adrift and Continents Aground* (San Francisco: W. H. Freeman, 1976).

16. Press and Siever, pp. 457–482.

17. Dietz and Holden.

Chapter 2

1. Charles Darwin, *The Descent of Man* (1871); reprinted in the Modern Library (New York: Random House), pp. 395–396.

2. Thomas H. Huxley, *Man's Place in Nature* (1863); reprinted as an Ann Arbor Paperback (Ann Arbor: University of Michigan Press), p. 123.

3. See S. L. Washburn, *The Study of Human Evolution*, Condon Lecture Series (Eugene, Oregon: University of Oregon Press, 1968).

4. Quoted in Huxley, p. 11.

5. Huxley, p. 36.

6. See Jane van Lawick-Goodall, *My Friends the Wild Chimpanzees* (Washington, D.C.: National Geographic Society, 1967); *In the Shadow of Man* (Boston: Houghton Mifflin, 1971). See also Jane van Lawick-Goodall and David A. Hamburg, "Chimpanzee Behavior as a Model for the Behavior of Early Man," in David A. Hamburg and H. Keith H. Brodie, eds., *American Handbook of Psychiatry*, vol. 6 (New York: Basic Books, 1975), pp. 14–43.

7. Goodall, *My Friends the Wild Chimpanzees*, p. 19.

8. George B. Schaller, *The Year of the Gorilla* (Chicago: University of Chicago Press, 1964), p. 201.

9. Phyllis C. Jay, ed., *Primates: Studies in Adaptation and Variability* (New York: Holt, Rinehart & Winston, 1968), pp. 487–488, 499, 501.

10. Jay, pp. 499.

11. Jay, p. 501.

12. Elwyn L. Simons, "The Earliest Apes," *Scientific American* 217 (December 1967), p. 35.

13. Louis S. B. Leakey, "Adventures in the Search of Man," *Scientific American* 208 (January 1963), p. 138.

14. Elwyn L. Simons, "Ramapithecus," *Scientific American* 236 (May 1977), pp. 28–35.

15. Simons, "Ramapithecus."

Chapter 3

Our gratitude to Kenneth Oakley for the chapter title, "Tools Makyth Man," Smithsonian Report (Washington, D.C., 1958).

1. K. R. L. Hall, "Tool-Using Performances as Indicators of Behavioral Adaptability," in Phyllis C. Jay, ed., *Primates: Studies in Adaptation and Variability* (New York: Holt, Rinehart & Winston, 1968), pp. 131–171.

2. Jane B. Lancaster, "On the Evolution of Tool-Using Behavior," *American Anthropologist* 70(1968), p. 58.

3. Lancaster, p. 62.

4. Charles Darwin, *The Descent of Man* (1871); reprinted in the Modern Library (New York: Random House), p. 435.

5. Darwin, p. 434.

6. Noel T. Boaz and John E. Cronin, ed., *Berkeley Papers in Physical Anthropology*, Kroeber Anthropological Society Papers, vol. 50 (Berkeley, Calif.: Kroeber Anthropological Society, 1977).

Chapter 4

1. Raymond A. Dart, *Adventures with the Missing Link* (New York: Harper & Brothers, 1959), p. 5.
2. Quoted in Dart, pp. 35–36.
3. Dart, pp. 49–50.
4. Quoted in Dart, p. 54.
5. Robert Broom, *Finding the Missing Link* (London: C. A. Watts & Company, 1950), p. 39.
6. Robert Broom, "The Ape-men," *Scientific American* 181 (November 1949), pp. 65–69.
7. W. E. Le Gros Clark, *Man-Apes or Ape-Men?* (New York: Holt, Rinehart & Winston, 1967).
8. W. E. Le Gros Clark, *Yearbook of Physical Anthropology* 5 (1949), p. 15. The Keith quotation appears in Dart, *Adventures with the Missing Link*, p. 81.
9. Quoted in Dart, p. 160.
10. Quoted in Dart, p. 162.
11. Louis S. B. Leakey, "Finding the World's Earliest Man," *National Geographic* 118 (September 1960), p. 425.
12. Personal communication to S. L. Washburn.
13. Le Gros Clark, *Man-Apes or Ape-Men?* pp. 48, 84.
14. F. Clark Howell, "Recent Advances in Human Evolutionary Studies," in S. L. Washburn and Phyllis Dolhinow, eds., *Perspectives on Human Evolution*, vol. 2 (New York: Holt, Rinehart & Winston, 1972), pp. 51–128.
15. D. C. Johanson and T. D. White, "A Systematic Assessment of Early African Hominids," *Science* 202 (1978), pp. 321–330.
16. Richard E. Leakey and Roger Lewin, *People of the Lake: Mankind and Its Beginnings* (New York: Doubleday, 1978), p. 122.

Chapter 5

1. See John Napier, "The Evolution of the Hand," *Scientific American* 207 (December 1962), pp. 56–62.
2. See Napier.
3. See Napier.
4. Glynn Isaac, "The Food-sharing Behavior of Protohuman Hominids," in Glynn Isaac and Richard E. Leakey, eds., *Human Ancestors* (San Francisco: W. H. Freeman and Company, 1979), pp. 110–123.
5. Noel T. Boaz and John E. Cronin, *Berkeley Papers in Physical Anthropology*, no. 50 (Berkeley, California: Kroeber Anthropological Society, 1977), p. 133.
6. Jane Goodall, "Life and Death at Gombe," *National Geographic* 155 (May 1979), pp. 592–620.
7. Richard B. Lee and Irven DeVore, eds., *Man the Hunter* (Chicago: Aldine, 1968).
8. See S. L. Washburn and C. S. Lancaster, "The Evolution of Hunting," in Lee and DeVore, pp. 293–303.
9. See Washburn and Lancaster.

10. S. L. Washburn, Phyllis C. Jay, and Jane B. Lancaster, "Field Studies of Old World Monkeys and Apes," *Science* 150 (1967), p. 1545.

11. E. Sue Savage-Rumbaugh and Beverly J. Wilkerson, "Socio-sexual Behavior in *Pan paniscus* and *Pan troglodytes:* A Comparative Study," *Journal of Human Evolution* 7 (1978), pp. 327–344.

12. W. B. Lemmon and M. L. Allen, "Continual Sexual Receptivity in the Female Chimpanzee," *Folia Primatologica* 30 (1978), pp. 80–88.

13. Ronald D. Nadler, "Determination of Sexuality in the Great Apes," in *Proceedings of the Workshop on the Conservation of the Orang Utan* (L. E. N. de Boev, ed.), Series Biogeographica, Dr. Junk International Publisher, Rotterdam (in press).

14. Personal communication to S. L. Washburn.

15. Andrew P. Wilson and R. P. Boelkins, "Evidence for Seasonal Variation in Aggressive Behavior by Macaques," *Animal Behavior* 18 (1970), pp. 719–724.

16. David A. Hamburg, "Emotions in the Perspective of Human Evolution," in Peter H. Knapp, ed., *Expression of the Emotions in Man* (New York: International Universities Press, 1963), p. 313.

17. Personal communication to S. L. Washburn.

18. Goodall, "Life and Death at Gombe."

19. S. L. Washburn and David A. Hamburg, "Aggressive Behavior in Old World Monkeys and Apes," in Phyllis C. Jay, ed., *Primates* (New York: Holt, Rinehart & Winston, 1968), pp. 458–478.

Chapter 6

1. J. S. Weiner, "Charles Darwin's *Descent of Man*—After 100 Years," in M. H. Day, ed., *Human Evolution* (New York: Barnes & Noble, 1973), pp. 1–11. Quotation on pp. 4–5.

2. Charles Darwin, *The Descent of Man* (1871); reprinted in the Modern Library (New York: Random House), pp. 434–436.

3. J. S. Weiner, *The Piltdown Forgery* (London: Oxford University Press, 1955).

4. A. Smith Woodward, *The Earliest Englishman* (London: Watts and Company, 1948).

5. W. E. Le Gros Clark, *Man-Apes or Ape-Men?* (New York: Holt, Rinehart and Winston, 1967), p. 66.

6. Le Gros Clark, pp. 22, 31.

7. Elizabeth R. McCown, *Sex Differences: Cranial Anatomy of Old World Monkeys*, unpublished doctoral dissertation, University of California, Berkeley, 1978.

8. Adrienne L. Zihlman et al., "Pygmy Chimpanzee as a Possible Prototype for the Common Ancestor of Humans, Chimpanzees and Gorillas," *Nature* 275 (1978), pp. 744–745.

9. David H. Hubel and Torsten N. Wiesel, "Brain Mechanisms of Vision," *Scientific American* 241 (September 1979), pp. 150–162.

10. Ronald E. Myers, "Comparative Neurology of Vocalization and Speech: Proof of a Dichotomy," in S. L. Washburn and Elizabeth R. McCown, eds., *Human Evolution: Biosocial Perspectives,* Perspectives on Human Evolution Series, vol. 3 (Menlo Park, California: Benjamin/Cummings, 1978), pp. 59–75.

11. Norman Geschwind, "Specializations of the Human Brain," *Scientific American* 241 (September 1979), pp. 180–199.

12. R. W. Sperry, "Lateral Specialization in the Surgically Separated Hemispheres," in F. O. Schmitt and F. G. Worden, eds., *The Neurosciences: Third Study Program* (Cambridge, Mass.: Massachusetts Institute of Technology Press, 1974), pp. 5–19.

13. Michael S. Gazzaniga and Joseph E. LeDoux, *The Integrated Mind* (New York: Plenum Press, 1978).

14. Bryan W. Robinson, "Anatomical and Physiological Contrasts between Human and Other Primate Vocalizations," in S. L. Washburn and Phyllis Dolhinow, eds., *Perspectives on Human Evolution,* vol. 2 (New York: Holt, Rinehart & Winston, 1972), pp. 438–443.

15. David Premack, *Intelligence in Ape and Man* (New York: John Wiley & Sons, 1976).

16. David A. Hamburg, "Emotions in the Perspective of Human Evolution," in Peter H. Knapp, ed., *Expression of the Emotions in Man* (New York: International Universities Press, 1963) p. 316.

17. S. L. Washburn, "Evolution and Education," *Daedalus* 103 (1974), pp. 221–228. See also S. L. Washburn, "Beyond the Basics," *Principal* 57 (1977), pp. 33–38.

Index